PROTEIN BIOCHEMISTRY, SYNTHESIS, STRUCTURE AND CELLULAR FUNCTIONS

HEAT SHOCK PROTEINS

CLASSIFICATION, FUNCTIONS AND APPLICATIONS

PROTEIN BIOCHEMISTRY, SYNTHESIS, STRUCTURE AND CELLULAR FUNCTIONS

Additional books in this series can be found on Nova's website under the Series tab.

Additional e-books in this series can be found on Nova's website under the e-book tab.

CELL BIOLOGY RESEARCH PROGRESS

Additional books in this series can be found on Nova's website under the Series tab.

Additional e-books in this series can be found on Nova's website under the e-book tab.

PROTEIN BIOCHEMISTRY, SYNTHESIS, STRUCTURE AND CELLULAR FUNCTIONS

HEAT SHOCK PROTEINS

CLASSIFICATION, FUNCTIONS AND APPLICATIONS

SAAD USMANI
EDITOR

New York

For permission to use material from this book please contact us:
Telephone 631-231-7269; Fax 631-231-8175
Web Site: http://www.novapublishers.com

Library of Congress Cataloging-in-Publication Data

ISBN: 978-1-62417-571-8

Library of Congress Control Number: 2012954737

Published by Nova Science Publishers, Inc. † New York

Contents

Preface

Heat shock proteins (HSP) are, from an evolutionary standpoint, an ancient family of cellular proteins involved with folding and unfolding of cellular proteins. They are present in all living organism, both unicellular and multicellular. The HSPs are also the most abundant cellular proteins and are also charged with protecting the plethora of "worker" protein in both physiological and stress conditions. Since their discovery in the early 1960s, many HSP families have been described.. The HSP classification has been historically based on their molecular weight (measured in kilodaltons), such as HSP27, HSP60, HSP70, HSP90, etc. Over the last decade, major advances have been made in understanding the role of HSPs in cellular functions.

The current book gathers state-of-the-art data on selected topic within plant, animal and human biology as they relate to the role of HSP. It includes chapters that give an overview of HSP involvement in immunity, cancer development, reproductive function, neurologic function, invertebrate biology and marine biology. This book is an endeavors to educate the reader on the diverse functions of the HSP family.

In: Heat Shock Proteins
Editor: Saad Usmani
ISBN: 978-1-62417-571-8

Chapter I

Heat Shock Proteins and Immune Functions

Robert Binder*
Department of Immunology, University of Pittsburgh, Pittsburgh, PA, US

Abstract

Heat shock proteins have been well studied with regards to their role in protecting cells from damage upon exposure to a range of stressful conditions. Heat shock proteins perform this protective function primarily by chaperoning and re-folding abnormal proteins. A number of heat shock proteins also have a unique role in immunology, performing a variety of functions including chaperoning of antigenic peptides for MHC presentation, transfer of antigens to, and, activation of Antigen Presenting Cells, and control of expression of a number of innate immune receptors. Some of these roles are possible due to the existence of an HSP receptor, CD91, expressed on Antigen Presenting Cells. The immunological roles of HSPs are discussed in this chapter.

Introduction

Heat shock proteins (HSPs) were first discovered when they were observed to be robustly expressed in cells incubated at elevated temperatures [1]. A large body of work has shown that a general role for HSPs is to protect cells from stressful conditions by chaperoning client proteins and helping them to refold in their correct conformations [2]. These stressful conditions can include aberrant temperature, glucose deprivation, hypoxia, infection by pathogens, among others. In the past three decades, select HSPs have been shown to play significant roles in various aspects of the immune system. These roles are dependent on the ability of a group of HSPs to chaperone not only proteins but also peptides. In this chapter, we review the roles of HSPs in the immune system with respect to their peptide chaperone

* Department of Immunology, University of Pittsburgh, E1051 BST, 200 Lothrop Street, Pittsburgh, PA 15262, Email: rjb42@pitt.edu.

function, the HSP receptors, the effects of HSPs on immune cells, and the role of HSP in innate immune system regulation, disease etiology and immunotherapy.

The Discovery of HSP Immunobiology

The ability of tumor cells to elicit specific protective immunity in mice [3-5] led to a search for tumor-derived molecules that were immunogenic. A biochemical approach that was taken involved the systematic chromatographic separation of various tumor cell components and testing each component in a tumor rejection assay- the same assay used to define immunogenicity of the whole cell. This approach led to the discovery of gp96 as a "tumor-specific transplantation antigen" [6]. Thus, immunization of mice with tumor-derived gp96 protected the mice against a late challenge of live tumor cells. These observations were soon tested for a few other HSPs including hsp70 [7], hsp90 [7], calreticulin [8], hsp110 [9] and grp170 [9] and the phenomenon held true for these other HSPs as well. Interestingly, in all cases tested, HSP isolated from a similar but distinct tumor type or from normal tissue failed to protect mice from a tumor challenge in each system tested. A systematic analysis of these HSPs isolated from normal or tumor tissues of different origin failed to reveal differences in sequence or structure of the respective molecule [10]. It was then hypothesized that molecules associated with the HSP accounted for the specific immune responses. The puzzling observation was solved when it was discovered that treatment of tumor-derived hsp70 with ATP abrogated the ability of hsp70 to elicit immune responses [11, 12]. ATP is known to remove peptide substrates from hsp70. It is now known that the specificity of the immune response elicited by HSPs is derived from the peptides that the HSPs chaperone and not the HSP molecule per se [13]. As discussed below, the ability of HSPs to chaperone a wide array of peptides and proteins is a property of the HSPs that is paramount to its immunobiology. The immunogenicity of HSPs has been demonstrated in prophylactic and therapeutic systems with a wide range of tumor models in mice and rats and in clinical trials in patients with cancers and infectious disease. This chapter shall focus on these six HSPs in general with specific references to each where necessary.

Hsps Chaperone Peptides within Cells for Immunological Purposes

The HSPs are only the second family of proteins, besides the MHC, that have peptide binding properties with roles in immunology. An examination of the nature of peptides bound to HSPs show no sequence consensus, thus no preference for particular amino acid residues. The evidence for peptide binding by HSPs is contributed by structural [14-20], biochemical [13, 21-30], and immunological [7-9, 31-35] data. The peptides are reflective of the entire proteome of the cell from which the HSP is isolated. In several cases these peptides have been shown to result from proteasomal degradation of cellular proteins in the cytosol [22, 24,-26, 30], although additional proteases may be involved. In normal cells these peptides will be self peptides, but in tumors or in cells infected with a pathogen those peptides include tumor antigens [7-9, 21, 30, 35] and pathogen-derived peptides [23, 29, 32, 33] respectively. The same holds true for systems tested with MHC differences [34] or cells expressing model antigens [13, 18, 25, 26, 28]. The tumor antigens include peptides derived from mutations in self proteins, from over-expressed proteins or aberrantly expressed proteins. The peptides associate with HSPs with broad specificity and relatively low affinity [13, 36]. These properties are shared with MHC however the MHC is more restrictive in peptide binding since there is a requirement for anchor residues at particular positions in the peptide sequence. Such anchor residues have not been identified as a requirement for HSP binding.

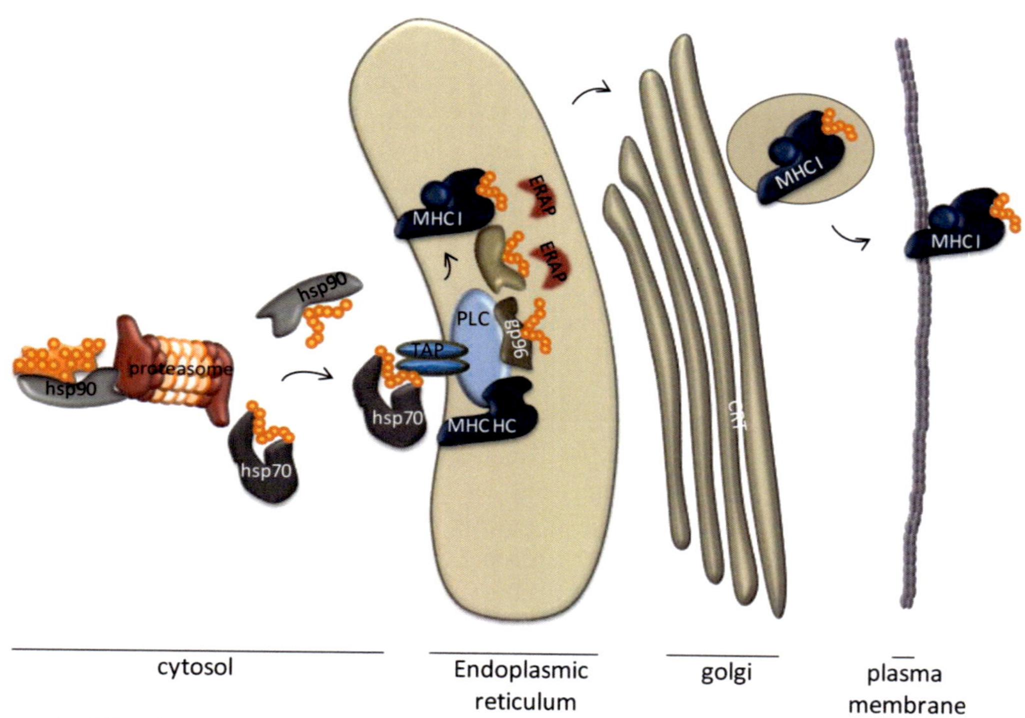

Figure 1. HSPs play a role in antigen processing for MHC I presentation. Proteins in the cytosol are shuttled by hsp90 to the proteasome for degradation. Peptides produced by the protesome are bound by hsp70 and hsp90 which transfer the peptides to TAP. TAP translocates peptides across the membrane into the ER. Here, gp96 and calreticulin sequentially transfer the peptides for loading onto the MHC. ERAP simultaneously trims peptides to the correct size. With the assistance of the PLC MHC are assembled with peptide and β2m. The mature MHC I traffics to the cell surgace for presentation to T cells.

As a chaperone of peptides, the HSPs have a direct role in antigen processing and MHC presentation (Figure 1). A pathway that includes the HSPs was proposed as early as 1994 [37], where the HSPs were suggested to form a relay line for transporting peptides. For MHC I the process of generating peptides for presentation begins in the cytosol with the large, multi-subunit catalytic protease called the proteasome [38]. The proteasome makes a series of digestions to convert proteins into peptides generally smaller than 20 residues. These peptides are direct substrates for the cytosolic hsp90 and hsp70 which chaperone the peptides to the transporter associated with antigen processing (TAP) [27, 28, 31]. The role for hsp90 and hsp70 in antigen processing and presentation, first proposed as a relay line of peptide transfer, has now been experimentally demonstrated in numerous studies [22, 27, 28, 31]. Inhibiting the association of peptides with either hsp70 [22] or hsp90 [28, 31] abrogates MHC I peptide presentation; this is a similar outcome to blocking the proteasome activity or TAP transport. The chaperoning of peptides within the cytosol by HSPs also satisfies the calculated efficiency of peptide production and MHC presentation which is much higher than could be sustained by random diffusion of naked peptides[39]. Indirect evidence is also provided by the physical interactions of hsp90 and hsp70 with the proteasome [40-42] and with TAP [43, 44], allowing for transfer of peptides between these molecules. TAP transports peptides into the ER in an ATP dependent manner. Once in the ER the peptides are loaded onto MHC heavy chains and assembled with β2-microglobulin into a tri-molecular complex with the assistance of the peptide loading complex (PLC) [38]. The PLC comprises of a number of proteins that includes the HSPs, calreticulin and gp96. There is evidence of sequential (linear) transfer of peptides between TAP, gp96, calreticulin, and MHC I [45]. Again the physical association of gp96 and calreticulin with MHC I in the ER lend support to their co-operative functions of peptide transfer [38, 46]. The other proteins of the PLC are largely involved in stabilization of the MHC heavy chain with the exception of ERAAP which trims peptides to their final length as they are simultaneously loaded onto MHC I [47, 48, 49]. The trimolecular complex is then transported to the cells surface for peptide presentation to T cells. Thus there is a significant role for selected HSPs in antigen processing and presentation for MHC I and cross-presentation. An in-depth role for HSPs in MHC II presentation has not as yet been clearly delineated.

The reason why the HSPs chaperone peptides within the cell is unlikely to have originated because of a necessity for MHC functions, since this property is an evolutionarily ancient property of HSPs, far predating adaptive immunity. Peptides of the kind chaperoned by HSPs for MHC presentation are typically too short to sustain higher order structures and due to hydrophobic constraints, require chaperones to shield hydrophobic residues. Indeed, a robust search for free peptides in cells has come to naught [50] and the HSPs may have a general role in chaperoning peptides for non-immunological functions as well. These functions are assumed to occur concurrently with chaperoning of other client molecules, whole proteins, as described below.

CD91 Serves as a Receptor for the Immunogenic HSPs

The immunogenic HSPs are abundant proteins comprising a large percentage of total cellular protein. The occurrence of these HSPs in the extracellular environment would suggest to immunological surveillance systems the presence of aberrant cells. The aberrant cells can include several types; (i) Pathogen infected cells- viruses or bacteria of the lytic type causing cell lysis by necrosis. The failing of cellular plasma membranes leads to the release of immunogenic HSPs. These HSPs, by virtue of their presence in the pathogen infected cell will chaperone pathogen-derived peptides as part of the total peptide repertoire. In several experimental systems, pathogen derived peptides have been isolated from HSPs. (ii) Tumors- as a result of hypoxia and other unfavorable conditions in tumor microenvironments, tumor cells may also lyse by necrosis releasing HSPs. In addition, several mechanisms of active release of HSPs by tumor cells have been described [51]. Regardless of the mechanism by which HSPs are released, the HSPs chaperone tumor specific peptides (antigens) in addition to normal/self peptides. The antigenicity of each tumor cell will thus be reflected in the peptide repertoire chaperoned by the HSP. As described above, this has been demonstrated in several murine and human tumor systems where tumor specific peptides have been isolated from HSPs. (iii) Trauma- abrupt disruption of cellular integrity, in a non-pathological condition, will release HSPs into the microenvironment. Here the peptide repertoire chaperoned by HSPs would be normal/self peptides.

Table 1.

Year (reference)	HSP interacting with CD91*	Type of evidence	Summary of the data
2000 (55)	gp96	Structural Immunological Biochemical	Physical interaction of CD91 with gp96. Blocking of cross-presentation of HSP-chaperoned peptides with CD91 antibodies
2001 (120)	calreticulin	Structural immunological	Calreticulin and CD91 interacted on the cell surface for uptake of apoptotic cells. Inhibition achieved with antibodies to either.
2001 (56)	gp96,hsp70, hsp90	Biochemical immunological	Expression of CD91 correlated with ability of APCs to cross-present. Cross-presentation of HSP-chaperoned peptides competed with other CD91 ligands.
2002 (121)	gp96	Biochemical structural	Blocking of gp96 binding with CD91 antibodies
2002 (122)	hsp70	Biochemical	Blocking of hsp70 binding with other CD91 ligands
2002 (123)	calreticulin	Biochemical immunological	Interaction of Calreticulin and CD91 is necessary for apoptotic cell ingestion by phagocytes.
2003 (124)	calreticulin	Biochemical immunological	Interaction of Calreticulin and CD91 is necessary to bind lung collectins which enhances inflammation.
2003 (83)	hsp70	Biochemical immunological	Binding of HSP70 to myeloid DC competitively inhibited by excess other CD91 ligands
2003 (125,126)	calreticulin	Biochemical immunological	Binding of recombinant Calreticulin and purified CD91 by co-immunoprecipitation
2003 (114)	Tumor lysate HSPs (Human)	Biochemical immunological	CD91 antibody prevented DC maturation and activation by HSPs.
2004 (57)	2004	Immunological Genetic biochemical	Competition for binding between gp96 and other ligands of CD91. Other CD91 ligands abrogated cross-presentation of gp96 chaperoned peptides. siRNA to CD91 abrogated binding and crosspresentation. CD91 antibodies prevented immune responses elicited by gp96.
2004 (127)	hsp70 (mycobacterial)	Biochemical immunological	CD91 antibody inhibited processing and presentation of extended hsp70 chaperoned peptides
2005 (128)	rhsp70 (mycobacterial/bovine)		Ligands of CD91 abrogated uptake of hsp70 by macrophages
2005 (129)	hsp70 (mycobacterial)	Biochemical immunological	CD91 antibodies inhibited processing and presentation peptides chaperoned by hsp70
2005 (130)	calreticulin		CD91 and calreticulin interaction is necessary for apoptotic cell engulfment
2007 (131)	calreticulin	Biochemical immunological	CD91 deficient PEA-13 cells did not bind CRT Binding of shared epitope (SE) blocked by anti-CRT and anti-CD91 antibodies and prevented DNA damage
2007 (132)	hsp90 homologue	Biochemical immunological	Antibodies or siRNA to CD91 blocked binding
2008 (133)	Hsp90	Biochemical genetic immunological	hsp90 induced cellular migration blocked by siRNA and antibodies to CD91
2008 (134)	hsp90 (frog)	Biochemical immunological	gp96 binding to cells blocked by CD91 antibodies

Table 1. (Continued)

2008 (135)	gp96	Biochemical immunological	gp96 binding to pDC inhibited by other CD91 ligands or antibodies to CD91
2009 (136)	Hsp90	Biochemical genetic immunological	knock down of CD91 abrogated gp96-mediated migration of cells and can be restored by CD91 cDNA transfection
2010 (67)	Gp96	Biochemical immunological	CD91 antibodies inhibited MHC II presentation of gp96-chaperoned peptides.
2010 (137)	Hsp70(human)	Biochemical genetic immunological	CD4+ T cell expansion, peptide cross-presentation, and expression of Th1/Th2 cytokines, induced by hsp70 treatment inhibited by siRNA to CD91
2010 (138)	Hsp90	Biochemical genetic immunological	Proximity ligation assays shows association of CD91 and hsp90 on HCT-8 cells. Hsp90 induced cell invasiveness decreased with antibodies or siRNA to CD91
2011 (139)	Hsp90	Biochemical immunological	Antibodies to HSP90 or CD91 blocked tumor invasion. Downstream signaling induced by HSP90 inhibited by siRNA to CD91
2011 (140)	Hsp70 (human)	Biochemical immunological	Other ligands or antibodies to CD91 inhibited proliferation of antiviral T cells from human samples
2011 (58)	Hsp70, hsp90,gp96	Biochemical genetic immunological	Hsp mediated activation of DCs inhibited with mutations in CD91 or antibodies to CD91.

*tested in a murine system except where noted in parenthesis.

In innate and adaptive immunological systems, the ability of hosts to defend against pathogens is a result of detection and response. Detection is through a set of receptors, the pathogen recognition receptors (PRRs) and the B cell and T cell receptors, respectively. There is a prerequisite that the receptors distinguish between self and foreign molecules and there are numerous mechanisms through which the receptors achieve this requirement. We have proposed a similar mechanism of detection and response that pertains to detection of abnormality through recognition of intracellular molecules in the extracellular environment [52]. As described below, this mechanism is important for initiation of priming adaptive immune responses. Several receptors that recognize intracellular proteins have been described. Receptor for Advanced Glycation End products (RAGE) has been described as a receptor for High Mobility Group Protein B1 (HMGB1), a nuclear DNA-binding protein [53]. Several molecules are suggested to be HSP receptors including CD91, SRA, SREC, and TLR2/4 [54]. In our view, a comprehensive argument has been made for CD91 as an HSP receptor and not the other molecules and so CD91 will be the focus here (Table 1). In some circles, the ligands of these receptors have collectively been called Danger Associated Molecular Patterns (DAMPs); a term analogous to Pathogen Associated Molecular Patterns (PAMPs) which bind to PRRs. CD91 is a large multi-domain receptor belonging to the class of scavenger receptors. It has been shown to bind to gp96, hsp70, hsp90 and calreticulin, thereby exerting numerous downstream effector responses [55-58]. The exact structural dynamics of how the different HSPs interact with CD91 has not been elucidated but it is envisaged that the HSPs may interact with different ligand binding domains of CD91. The downstream immunological responses of CD91 engagement by HSPs are at least two fold-cross-presentation of the HSP-chaperoned peptide [55-57] and stimulation of the antigen presenting cell to provide co-stimulation [58]. Both of these effects elicited by HSPs are abrogated with a genetic loss of CD91 [57], CD91 competing ligands or mutations in residues within the signaling motifs of CD91 or are absent in CD91 non-expressing cells [58].

Extracellular HSPs Provide Two Signals Necessary for T Cell Priming (Figure 2)

Cross-presentation refers the acquisition of antigens by antigen presenting cells (APCs) and the processing and presentation of these antigen, as peptides, by MHC I molecules. This is a unique feature of a select subset of antigen presenting cells called dendritic cells (DC). In some systems, the cross-presenting dendritic cell can further be subtyped into CD8a^{+} or CD103^{+} DC for experimental models involving model antigens and viral recrudescence respectively [59, 60]. With respect to the HSPs, CD91 has been shown to be the major receptor involved in cross-presentation of HSP-chaperoned peptides [55-57]. Non-CD91 binding proteins often used as a control, such as serum albumin, are unable to channel their chaperoned peptides into the cross-presentation pathway. CD91-dependent endocytosis of HSP-peptide complexes has been observed with the complex ending up in endosomes. Through chemical inhibition of proteasome or vesicular traffic the processing and trafficking of the peptide has been followed [56]. A remaining question pertains to how the peptide, by itself or in conjunction with the HSP cross endosomal membranes into the cytosol for processing by the proteasome. Of the several alternative mechanisms put forward, one has been described which involves a role for endogenous hsp90 [61-63]. In this mechanism, hsp90 is shown to physically associate with the antigen being cross-presented and shuttles it across the membrane into the cytosol for further processing by the proteasome. As shown in numerous antigenic systems, the ultimate effect is that the peptide, regardless of the initial length, originally chaperoned by the HSP is presented by MHC I. Peptides chaperoned by HSPs can also be presented by MHC II molecules in a separate pathway [64-67] that is also CD91 dependent [67]. The HSP-CD91 pathway provides an efficient mechanism for the transfer of antigens from the antigen bearing cell to the APC for the purposes of priming T cell responses. An examination of the quantity of the antigen made in a tumor cell, for example, is 50,000 times less than is required for sufficient cross-presentation, if the antigen as a whole protein by itself were being acquired by the APC [68, 69]. A simple test for the requirement of HSPs for tumor antigen cross-presentation was performed by depleting tumor cells of the major immunogenic HSPs. HSP-depleted tumor cells were unable to cross-present antigens even though they still had the intact antigen (whole protein) [69]. In a second system, tumor cells expressing molecules that abrogate binding of HSP to CD91, such as the receptor associated protein (RAP), decreased the immunogenicity of the tumor cells, in other words rendering the tumor cells less capable of priming immune responses when compared to non-RAP expressing parental tumors [52]. The differences were absent when tested in immunocompromised mice. The conclusion from these studies was that HSPs are necessary and sufficient for cross-presentation of tumor antigens. Other studies corroborate this [70-72]. In other systems where antigen load is orders of magnitude larger, such as in some viral infections, other mechanisms may be employed that are not necessarily dependent on the HSP-CD91 pathway [73]. Peptide presentation by MHC molecules is one of at least two signals required for priming T cell responses.

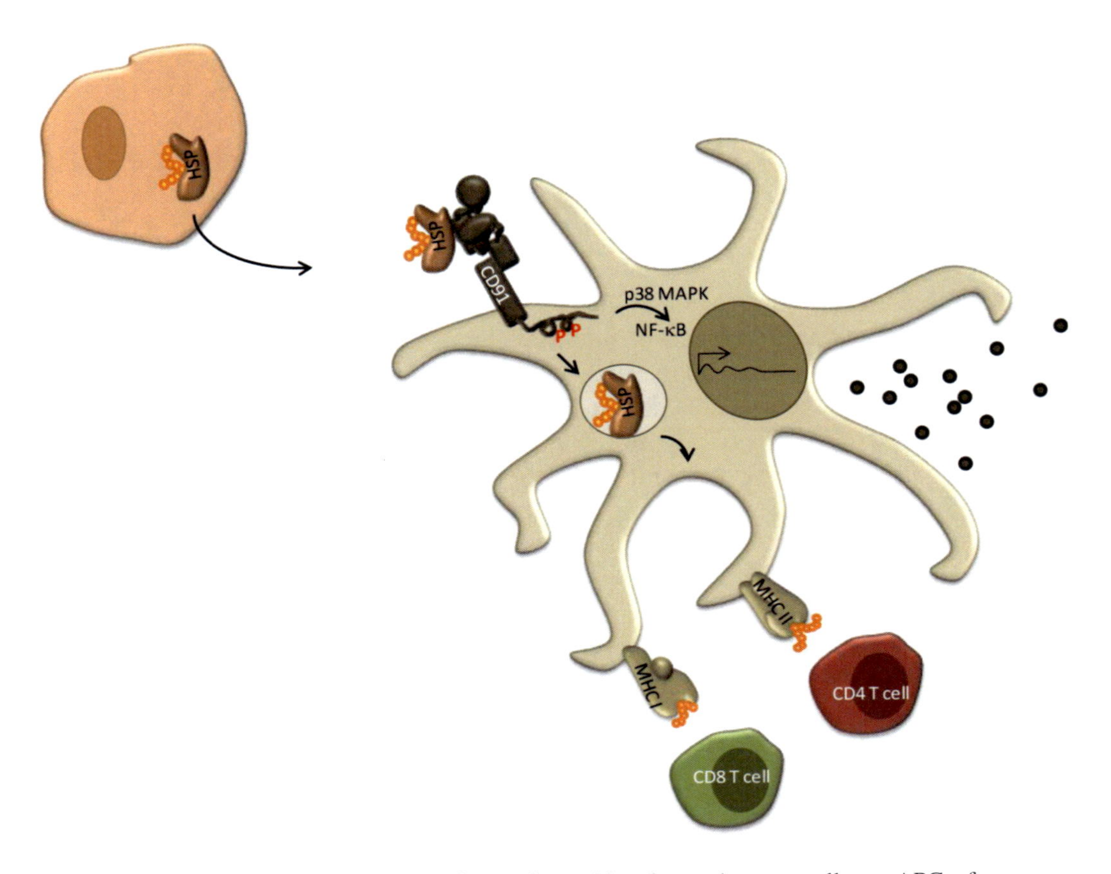

Figure 2. HSPs transfer antigens in the form of peptides from aberrant cells to APCs for cross-priming.HSPs released from aberrant cells through a variety of mechanisms interact with CD91 on the surface of APCs. CD91 performs 2 roles; first, it internalizes the HSP-peptide complex and peptide is processed and presented on either MHC I or MHC II, and second, CD91 is phosphorylated on the cytoplasmic tail and transmits intracellular signals to the nucleus allowing the APC to mature and present co-stimutation in the form of unique patterns of cytokines and co-stimulatory molecules. These two events allow for priming of specific T cell responses.

A second signal required for T cell priming is co-stimulation that is provided by the APC in conjunction with the MHC-peptide complex on the cell surface. Co-stimulation is a combination of ligand-receptor interactions and cytokines and the specific set of co-stimulation that is provided by the APCs determines the type of T cell response that is primed. HSPs, independent of bound peptide, are able to stimulate APCs, enabling them to provide co-stimulation. A number of cytokines are released by HSP-stimulated APCs including TNF-α, IL-1β, IL-6, IL-12, GM-CSF [58, 74-77]. In addition, HSP-stimulated APCs up-regulated expression of co-stimulatory molecules including CD80, CD86, CD40 and MHC II[74]. Although this is not an exhaustive list of co-stimulatory molecules and cytokines, the complete profile is dependent on the type of APC (macrophage or DC) that is stimulated and the HSP (hsp70, hsp90, calreticulin or gp96) that is used for stimulation. In one study, primary APCs were shown to be activated by HSPs in a CD91-dependent manner initially suggesting that CD91 was acting as a signaling receptor for the immunogenic HSPs [58]. The β-chain of CD91 has two NPXY motifs which are consensus motif for phosphorylation and signal transduction. Upon mutation of the tyrosines to phenylalanine, CD91 failed to transmit signals in response to HSP stimulation, abrogating the co-stimulation provided by the APC. The signaling pathway(s) initiated by CD91 upon HSP stimulation involves the activation of NF-kB [58], 74 and p38 MAPK [58] although other molecules are yet to be identified. These pathways are HSP specific and are dependent on the type of APC exposed to the HSP.

CD91 thus has a role in signal 1 (cross-presentation) and 2 (co-stimulation) that is provided by the APC to T cells in response to extracellular HSP. These results have implications in several fields of immunology described below.

Importance of Extracellular HSPs to Concomitant Immunity, Autoimmunity and Inflammation

Concomitant immunity was first described over a hundred years ago [78, 79] and defines the priming of tumor specific immune responses in the tumor bearing host. Therefore, the growth of a second tumor is regressive when implanted in a mouse already bearing a prior established tumor. With the advent of syngeneic mice, the phenomenon was confirmed in studies by North and colleagues in the 1970s [80]. Several different tumor systems were used to show the generality of the observation. Even though those early experiments were done with transplantable tumors in mice, the existence of tumor specific T cells has been demonstrated in a vast number of patients with a variety of (spontaneous) cancer types[81]. To date there is no answer as to how the priming event happens. Although tumors express antigens due to mutations, abnormal expression patterns or aberrant post-translational modifications, those antigens are generally quantitatively insufficient for cross-priming [68,69] and, in general, do not provide the co-stimulation necessary for priming T cell responses. These two quagmires are satisfied by the immunogenicity of HSPs. First, HSPs increase the efficiency of cross-presentation by several orders of magnitude by transferring the antigen in the form of peptides to APCs and thus T cells can be primed against very low antigen levels [69]. These studies have been discussed in context of other proposed forms of antigen transfer; vesicular bodies such as apoptotic bodies are generally deemed tolerogenic, while evidence for a role for exosomes *in vivo* is largely lacking. Secondly, HSPs activate APCs to provide co-stimulation [58, 74-77] and were the first so called danger associated molecular patterns (DAMPs) molecules to be discovered. A note of other self molecules that can act as DAMPs such as dsDNA [82] and HMGB1 [53] is made however those molecules do not provide the antigen necessary for the specificity of the immune response. We opine that the provision of 'danger' and antigen occurs as one unit- the HSP-peptide complex. This opinion is supported by emerging experimental data [52].

The role of HSPs in autoimmune diseases continues to be studied. Although the etiology of rheumatoid arthritis remains unresolved, several factors that contribute to the initiation and/or progression of the disease can be pinpointed. The observation of elevated levels of hsp70 in synovial fluids from inflamed joints of RA patients is one of these factors. Hsp70 is found both within the fibroblasts at the joint and in the fluid itself [83, 84]. As mentioned above, hsp70 can interact with its cell surface receptor CD91, and potentially other receptors, on cells to induce the release of pro-inflammatory cytokines such as IL-1β, IL-6 and TNF-α

among others. The correlation of elevated hsp70 levels in synovial fluids from inflamed joints compared to non-inflamed joints implicates hsp70 as an initiator of inflammation and/or a perpetrator of these events. In this disease hsp70 will chaperone self-peptides that can be cross-presented by local APCs [85]. This cross-presentation of self-antigens appears to be sufficient to break tolerance and for priming self antigen-specific T cells which could contribute to cellular destruction observed in arthritic joints.

HSPs as Chaperones of Client Proteins with Immunological Functions

HSPs are widely known for their ability to chaperone proteins. As one of their major roles, HSPs in the ER bind to nascent polypeptides to enable the protein to fold into thermodynamically stable, higher order conformations. The HSPs appear to bind peptides and proteins in different ways and it is not known if the two chaperone functions occur sequentially or simultaneously. In the recent years a number of proteins with important immunological functions have been found to be dependent on the HSP, gp96, for proper expression.

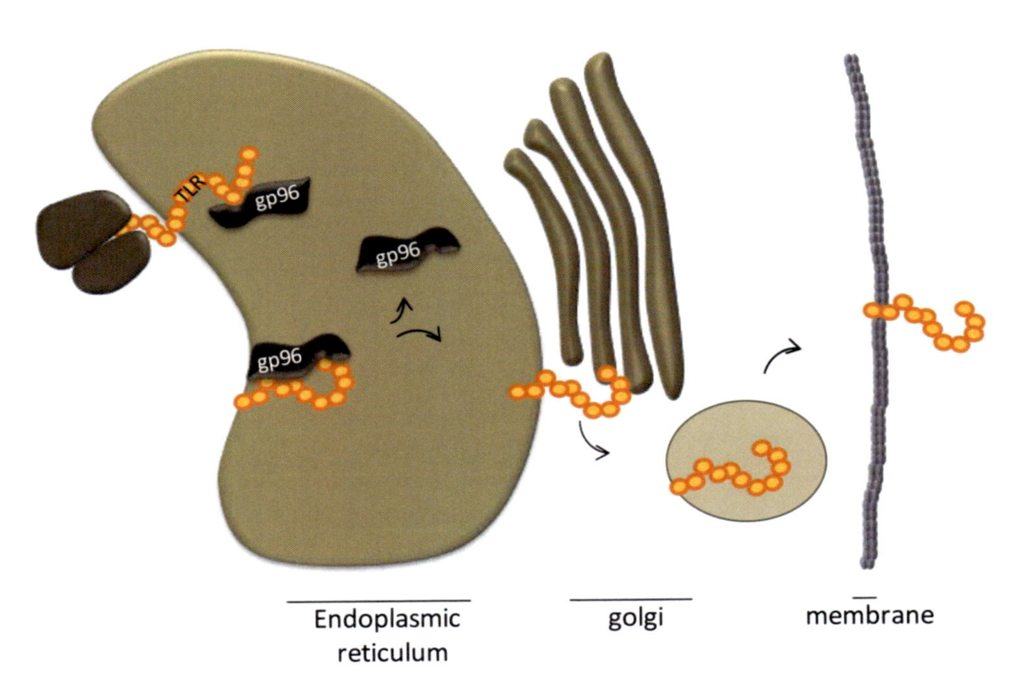

Figure 3. Gp96 is involved in the expression of immunologically important molecules. Client proteins including the TLRs are translated into the ER by ribosimes. The nascent polypeptide chains are bound by gp96 and are folded into correct spatial conformations. The fully matured client protein then traffics to the membranes and gp96 is retained in the ER.

1) Toll like receptors (TLRs) and integrins (Figure 3): TLRs are important for recognition of various pathogen associated molecular patterns (PAMPs) and initiate innate immune

responses by triggering the cell to release a number of pro-inflammatory cytokines [86]. The TLRs that bind to bacterial products (TLR 1, 2, 4, 5, 6) are expressed on the plasma membrane while those that bind viral products (TLR 3, 7, 8, and 9) are largely expressed in endosomes. A few TLRs are also known to homo- or hetero- dimerize to form functional receptors (TLR2/6, TLR). The difference in cellular localization and quaternary structure (or lack of) highlights potential differences in post translational events prior to full expression. Despite the differences in cellular localization and/or dimerization, gp96 has been found to be a master regulator of the expression of TLR2, TLR4, TLR5, TLR7, and TLR9 [87]. A loss of gp96 expression in macrophages eliminated the expression of these TLRs and thus rendered the cells non-responsive to their respective ligands [87]. Mice, lacking gp96 expression specifically in macrophages, were further shown to be resistant to endotoxin shock and were highly susceptible to Listeria monocytogenes infection. Further investigation has shown that TLRs interact with the client binding domain at the C-terminal hydrophobic region of gp96. This region is also responsible for binding a number of other client proteins such as select intergrins, however subtle differences exist in the way each client (TLR or integrin) interacts with gp96 [88]. Due to its role in chaperoning TLR and integrins for optimal expression, gp96 has a regulatory role in hematopoiesis. A loss of gp96 expression in hematopoietic cells leads to a severe defect in development of B and T cells [89].

2) LDL receptor family members: A comprehensive study was performed on plasma membrane proteins to determine which proteins were expressed in a gp96 dependent manner. By profiling plasma membrane proteins of gp96-deficient or gp96-reconstituted cells, over a 100 proteins were identified to be differentially expressed [90]. Interestingly four members of the LDL receptor family were identified; LDLR, LRP6, Sorl1 and LRP8. These proteins share a common β-propeller and leucine-rich repeat, which appear to be co-existing common structural motifs necessary for binding gp96. Interestingly, the HSP receptor CD91 (LRP1) which has these same structural motifs was not found and there may be other structural requirements for binding. Evidently not all proteins that have the propeller domain motifs require gp96 for correct folding; however the unique boundaries between the propeller motif and adjacent motifs may require gp96 during assembly. The relevance of one LDL receptor family member, CD91, to the immune response has been highlighted above.

3) Oncogenes and tumor antigens: As a chaperone of proteins within the cytosol, hsp90 serves to refold proteins. The absence of hsp90 chaperone function leads to the ubiquitin-mediated degradation of potential client proteins by the proteasome. Thus oncogenes or their products that are clients of hsp90 are of particular interest since inhibition of hsp90 leads to their degradation. Several of these oncogenes such as HER2, ALK, EGFR, and BRAF are known to require hsp90 for their expression. Small molecules such as geldanamycin and its derivatives, radicicol and taxol are all being utilized for hsp90 inhibition with the purpose of abrogating cancer growth. Clinical trials with these drugs are being tested in patients with cancers that are dependent on the presence of or driven by particular oncogene (product) [91].

In addition to degradation of particular oncogenes and other proteins, hsp90 inhibition has been shown to increase MHC I presentation of peptide products of the degradation. The increased presentation of peptides makes the cells better targets for T cell recognition and priming. In one example, inhibition of hsp90 in cancer cells led to the increased degradation of EphA2 and better recognition of those cancer cells by EphA2 specific T cells [92]. This increased T cell response has been verified in mouse models *in vivo* and can lead to tumor rejection. At first glance, this observation appears to be contradictory to the role of hsp90 in chaperoning peptides for MHC I presentation [28, 31]. It appears that the finely tuned concentrations of hsp90 inhibitors used in various cells can distinguish between the protein and peptide chaperoning activities of hsp90, with a low dose affecting the protein chaperoning activity and a ten-fold higher of the same drug inhibiting the peptide chaperoning activity. These numbers should be a very important consideration in determining dosages in clinical studies where hsp90 inhibitors are used to treat cancer patients.

Application of HSPs to Immunotherapy of Cancer, Infectious Disease and Autoimmunity

i. Cancer: In experimental models of cancer in rodents, treatment of tumor bearing animals with tumor-derived HSPs retarded the growth of primary tumors when compared to untreated or control treated mice. Furthermore, in metastatic tumor models (where the primary tumor was removed following establishment of lung metastasis), treatment of mice with HSPs led to significantly decreased number of metastatic nodules and prolonged survival of animals compared to control treated animals [35]. These observations comprised the preclinical data. There are three theoretical considerations in the clinical trials described below: (a) By definition, the HSP vaccine consists of the HSP molecule itself and the peptides it chaperones which are naturally associated with it after purification. As discussed above these peptides will typically reflect the entire antigenicity of the cell from which the HSP was purified. By its inherent nature, the vaccine must be produced from each patient's tumor. (b) During the first trials with autologous HSP vaccines, these trials had no precedent and thus the optimal doses of HSPs to administer to patients provided a unique obstacle. Studies over the years have shown that, unlike chemotherapy, maintaining particular blood levels of HSP is unnecessary. In mice and humans alike the receptor density and the APC localization and numbers are similar and so the doses used in mice should lie within similar ranges as those to be used in humans. Consistent with this idea is the observation that optimal doses of HSPs that provide immune responses in mice are the same as those used in rats for comparable responses even though there is a greater than 20-fold difference in body weight between the two rodents[7, 93]. Since the vaccine is isolated from the tumor the amount of tumor obtained after surgery determines the availability of the vaccine and dosage becomes an important factor. For 10 immunizations at 25μg per dose, 250μg is required and this can be obtained from 2-5g of tumor tissue. (c) In murine models of cancer, there are easy measurable end points to determine the efficacy of vaccines such as measurement of tumor size or survival of animals. Examination of immunological readouts tend to assume a second priority and they include, where applicable, CTL assays or tetramer staining for tumor specific T cells, Ig titers, and SEREX to mention a few. On the contrary, in the human disease, more emphasis has generally been placed on the immunological readouts usually at the expense of clinical response which is the true determinant of successful immunization. However, there is generally a lack of true surrogate markers because successful immunization is not necessarily reflected in distinct T cell responses as measured *in vitro* and distinct T cell responses are not necessarily indicative of protection from cancer [94-101].

The clinical trial with HSP-based vaccines began in 1995 with four preliminary trials [94, 101-105]. These Phase I/II trials are non-randomized and performed largely to establish dosages and address potential toxicities while also broadly comparing immune responses with those obtained in mice. The trials consisted of patients with renal cell carcinoma and melanoma. The data obtained from these trials were consistent with results obtained in mice; autologous HSP vaccine elicited powerful $CD4^+$ and $CD8^+$ T cell responses against the tumor and did not produce any toxicity. Clinical responses from these trials were only suggestive they could only be compared with historical controls. In addition, these trials demonstrated the feasibility of the manufacture of autologous HSP vaccines from each patient's own tumor. Other Phase I and II trials have either been completed in pancreatic cancer, colon cancer with similar outcomes or are ongoing in B lymphoma, CML, lung cancer and glioma [106, 107].

These earlier trials have been followed up by two phase III trial in patients with melanoma and renal cell carcinoma. The details of the design and early results from those trials have been published [108, 109]. The early analysis in these patients shows a trend in favor of the HSP vaccine compared to the observation arm (RCC) or patients receiving physician's choice of treatment (melanoma), however these data have not yet achieved statistical significance. Indeed, subset analysis in the RCC trial showed that in 361 patients with better prognostic factors (stage I/II high grade, III T1, T2, T3a, low grade classified as "intermediate risk"), a significant benefit was seen in the treatment arm [108]. This is also true in subset analysis in the melanoma trial [109]. Patients continue to be followed for overall survival. This represents the largest randomized study to date for renal cell carcinoma in the adjuvant setting. In all trials, the vaccine was well tolerated and adverse events reported were generally mild and expected and related to the actual injection.

ii. Infectious disease: We and others have routinely been able to prime pathogen specific Th1 responses in experimental animals [23, 29, 32, 33] and humans [110, 111] using HSPs chaperoning pathogen specific peptides. These are characterized by both CD8+ (CTL) and CD4+ T cell responses with specificity for peptides derived from the pathogen. The experimental systems include studies with viruses [23, 29, 32, 110-112] and bacteria [33]. In humans, immunization with hsp70 complexed with a set of HSV derived peptides led to priming of HSV specific CD8 and CD4 cells as determined by ELISPOT assays performed post immunization. Interestingly, in a few studies in mice, immunization with gp96 isolated from virus infected cells led to the development of anti-viral antibodies [32, 112]. There is currently no explanation why the Th responses are skewed either way in each immunization protocol. We have learned lessons from recent observations that additional cytokines released by (tumor) cells in the immediate environment of vaccination may complement, inhibit or supersede the ones released by HSP-stimulated APCs [58]. In the infectious disease models those cells would be the virally infected cells that are known to secrete IFN-α and IFN-β. We anticipate that further clinical trials will be initiated in patients with various infectious diseases. In another recent development, a strong correlation has been established between HIV infected patients classified as "true" long term non-progressors and CD91 expression [113-115]. The idea is that the strong expression of CD91 in these patients allows for more efficient cross-presentation of HIV derived peptides and priming of HIV-specific T cell responses [114]. These enhanced T cell responses account for the maintenance of low viral titers in these patients. Further studies on this subject should clarify the role of extracellular HSPs in this setting.

iii. Autoimmunity: Immunization of mice with HSPs typically elicits Th1 responses characterized by CTL specific for antigenic peptides chaperoned the HSP. However, higher doses of HSPs (10 times the immunizing dose) administered to mice have been shown to prime an immunosuppressive phenotype characterized by expansion of $CD4^+$ Treg [116-118]. The application of this “high dose’ phenomenon has been tested in mouse models of autoimmunity including diabetes and Experimental autoimmune encephalomyelitis [117]. In those models, administration of “high doses” of HSPs reduced the severity of disease or prevented its development outright. In addition high doses of HSPs can significantly delay the rejection of minor and/or major histocompatible complex mismatched skin grafts [118]. These effects are independent of specific peptides chaperoned by the HSP. While the phenomenon is well established the mechanism has many unanswered questions. Higher doses of HSPs may target different subsets of APCs or stimulate the same APCs (as the immunogenic dose) to elicit distinct co-stimulatory profiles that will prime the observed Treg cell expansion [119]. This remains to be experimentally demonstrated. Definition of these mechanisms will offer targets for inhibition for therapy of autoimmune diseases.

Conclusion

Over the past 3 decades the various roles of HSPs in the immune systems have been explored and characterized. It appears that the evolutionarily ancient chaperone functions of HSPs in binding peptides and proteins have been commandeered by the relatively recent development of adaptive immune system. However recent studies suggest that parallel to evolution of innate responses, multicellular organisms are alerted to aberrant cellular damage by utilization of pre-existing receptors (CD91) to detect the presence of abundant intracellular molecules (HSPs). We draw many similarities between the innate immune responses elicited by PAMPs through PRRs and those by HSPs via CD91 in terms of co-stimulation for T cell priming. Indeed the HSP-CD91 network has been well documented not only in mammals but also in amphibians. While CD91 is a well documented receptor for HSPs (Table 1), there may be other molecules that may serve as receptors, offering a diversity of responses that may be elicited by each HSP. Again, the diversity of PRRs for recognition of various PAMPs is well noted in innate immunity. The necessary role of HSPs in expression of PRRs (TLRs) strongly suggests that the two parallel pathways may regulate each other and this offers new opportunities for exploration. The responses observed with simultaneous release of multiple HSPs from a cell have not been studied and may offer new surprises. The immunogenicity of HSPs has been harnessed for clinical benefit and those trials are ongoing. With greater understanding of the immunobiology of these proteins, we anticipate that vaccine design will be enhanced. Finally, we offer a cautionary note to clinical trials where HSPs themselves are targeted by chemotherapeutic agents to inhibit their chaperone function.

Acknowledgments

I would like to acknowledge the members of my laboratory Dr. Sudesh Pawaria, Ms. Michelle Messmer, Mr. Yu Jerry Zhou and Ms. Leticia Monin for critically reading the chapter. Our work cited in this chapter is funded by NIH grants CA137133 and AI079057 awarded to RB.

References

[1] Ritossa F. A new puffing pattern induced by temperature shock and DNP in drosophila. *Experientia.*18, 571-573, (1962).

[2] Lindquist S, Craig EA. The heat-shock proteins. *Annu Rev Genet.* 22, 631-677, (1988)

[3] Klein G, Sjogren HO, Klein E, Hellstrom KE. Demonstration of resistance against methylcholanthrene-induced sarcomas in the primary autochthonous host. *Cancer Res.* 20, 1561-1572, (1960).

[4] Prehn RT, Main JM. Immunity to methylcholanthrene-induced sarcomas. *J Natl Cancer Inst.* 18, 769-778, (1957).

[5] Basombrío MA. Search for common antigenicities among twenty-five sarcomas induced by methylcholanthrene. *Cancer Res.* 30, 2458-2462, (1970).

[6] Srivastava PK, DeLeo AB, Old LJ. Tumor rejection antigens of chemically induced sarcomas of inbred mice. *PNAS U S A.* 83, 3407-3411 (1986).

[7] Udono H, Srivastava PK. Comparison of tumor-specific immunogenicities of stress-induced proteins gp96, hsp90, and hsp70. *J Immunol.* 152, 5398-5403 (1994).

[8] Basu S, Srivastava PK. Calreticulin, a peptide-binding chaperone of the endoplasmic reticulum, elicits tumor- and peptide-specific immunity. *J Exp Med.* 189, 797-802 (1999).

[9] Wang XY, Kazim L, Repasky EA, Subjeck JR. Characterization of heat shock protein 110 and glucose-regulated protein 170 as cancer vaccines and the effect of fever-range hyperthermia on vaccine activity. *J Immunol.* 166, 490-497 (2001).

[10] Srivastava P.K. Peptide-binding heat shock proteins in the endoplasmic reticulum: role in immune response to cancer and in antigen presentation. *Adv Cancer Res.* 62, 153-177 (1993).

[11] Li Z, Srivastava PK. Tumor rejection antigen gp96/grp94 is an ATPase: implications for protein folding and antigen presentation. *EMBO J.* 12, 3143-3151 (1993).

[12] Srivastava PK, Heike M. Tumor-specific immunogenicity of stress-induced proteins: convergence of two evolutionary pathways of antigen presentation? *Semin Immunol.* 3, 57-64 (1991).

[13] Binder RJ, Blachere NE, Srivastava PK. Heat shock protein-chaperoned peptides but not free peptides introduced into the cytosol are presented efficiently by major histocompatibility complex I molecules. *J Biol Chem.* 276, 17163-17171 (2001).

[14] Zhu X, Zhao X, Burkholder WF *et al.* Structural analysis of substrate binding by the molecular chaperone DnaK. *Science.* 272, 1606-1614 (1996).

[15] Stebbins CE, Russo AA, Schneider C, Rosen N, Hartl FU, Pavletich NP. Crystal structure of an Hsp90-geldanamycin complex: targeting of a protein chaperone by an antitumor agent. *Cell.* 89, 239-250 (1997).

[16] Linderoth NA, Popowicz A, Sastry S. Identification of the peptide-binding site in the heat shock chaperone/tumor rejection antigen gp96 (Grp94). *J Biol Chem.* 275, 5472-5477 (2000).

[17] Gidalevitz T, Biswas C, Ding H *et al.* Identification of the N-terminal peptide binding site of glucose-regulated protein 94. *J Biol Chem.* 279, 16543-16552 (1994).

[18] Binder RJ, Kelly JB 3rd, Vatner RE, Srivastava PK. Specific Immunogenicity of Heat Shock Protein gp96 Derives from Chaperoned Antigenic Peptides and Not from Contaminating Proteins. *J Immunol.* 179, 7254-7261 (2007).

[19] Chouquet A, Païdassi H, Ling WL, Frachet P, Houen G, Arlaud GJ, Gaboriaud C. X-ray structure of the human calreticulin globular domain reveals a peptide-binding area and suggests a multi-molecular mechanism. *PLoS One.* 6, e17886, (2011).

[20] Stocki P, Morris NJ, Preisinger C, Wang XN, Kolch W, Multhoff G, Dickinson AM. Identification of potential HLA class I and class II epitope precursors associated with heat shock protein 70 (HSPA). *Cell Stress Chaperones.* 15, 729-741 (2010).

[21] Peng P, Ménoret A, Srivastava PK. Purification of immunogenic heat shock protein 70-peptide complexes by ADP-affinity chromatography. *J Immunol Methods.* 204, 13-21 (1997).

[22] Binder RJ, Blachere NE, Srivastava PK. Heat shock protein-chaperoned peptides but not free peptides introduced into the cytosol are presented efficiently by major histocompatibility complex I molecules. *J Biol Chem.* 276, 17163-17171 (2001).

[23] Nieland TJ, Tan MC, Monne-van Muijen M, Koning F, Kruisbeek AM, van Bleek GM. Isolation of an immunodominant viral peptide that is endogenously bound to the stress protein GP96/GRP94. *Proc Natl Acad Sci USA.* 93, 6135-6139 (1996).

[24] Heikema A, Agsteribbe E, Wilschut J, Huckriede A. Generation of heat shock protein-based vaccines by intracellular loading of gp96 with antigenic peptides. *Immunol Lett.* 57, 69-74 (1997).

[25] Breloer M, Marti T, Fleischer B, von Bonin A. Isolation of processed, H-2Kb-binding ovalbumin-derived peptides associated with the stress proteins HSP70 and gp96. *Eur J Immunol.* 28, 1016-1021 (1998).

[26] Arnold D, Wahl C, Faath S, Rammensee HG, Schild H. Influences of transporter associated with antigen processing (TAP) on the repertoire of peptides associated with the endoplasmic reticulum-resident stress protein gp96. *J Exp Med.* 186, 461-466 (1997).

[27] Ishii T, Udono H, Yamano T *et al.* Isolation of MHC class I-restricted tumor antigen peptide and its precursors associated with heat shock proteins hsp70, hsp90, and gp96. *J Immunol.* 162, 1303-1309 (1999).

[28] Kunisawa J, Shastri N. Hsp90alpha chaperones large C-terminally extended proteolytic intermediates in the MHC class I antigen processing pathway. *Immunity.* 24, 523-534 (2006).

[29] Meng SD, Gao T, Gao GF, Tien P. HBV-specific peptide associated with heat-shock protein gp96. *Lancet.* 357, 528-529 (2001).

[30] Castelli C, Ciupitu AM, Rini F *et al.* Human heat shock protein 70 peptide complexes specifically activate antimelanoma T cells. *Cancer Res.* 61, 222-227 (2001).

[31] Callahan MK, Garg M, Srivastava PK. Heat-shock protein 90 associates with N-terminal extended peptides and is required for direct and indirect antigen presentation. *Proc Natl Acad Sci U S A.* 105, 1662-1667 (2008).

[32] Navaratnam M, Deshpande MS, Hariharan MJ, Zatechka Jr DS, Srikumaran S. Heat shock protein-peptide complexes elicit cytotoxic T-lymphocyte and antibody responses specific for bovine herpesvirus 1. *Vaccine.*19, 1425-1434 (2001).

[33] Zugel U, Sponaas AM, Neckermann J, Schoel B, Kaufmann SH. 96-peptide vaccination of mice against intracellular bacteria. *Infect Immun.* 69, 4164-4167 (2001).

[34] Arnold D, Faath S, Rammensee H, Schild H. Cross-priming of minor histocompatibility antigen-specific cytotoxic T cells upon immunization with the heat shock protein gp96. *J Exp Med.* 182, 885-889 (1995).

[35] Tamura Y, Peng P, Liu K, Daou M, Srivastava, PK. Immunotherapy of tumors with autologous tumor-derived heat shock protein preparations. Science. 278, 117-120 (1997).

[36] Demine R, Walden P. Testing the role of gp96 as peptide chaperone in antigen processing. *J Biol Chem.* 280, 17573-17578, (2005).

[37] Srivastava PK, Udono H, Blachere NE, Li Z. Heat shock proteins transfer peptides during antigen processing and CTL priming. *Immunogenetics.* 39, 93-98, (1994).

[38] Cresswell P, Ackerman AL, Giodini A, Peaper DR, Wearsch PA. Mechanisms of MHC class I-restricted antigen processing and cross-presentation. *Immunol Rev.* 207, 145-157, (2005).

[39] Yewdell JW. Not such a dismal science: the economics of protein synthesis, folding, degradation and antigen processing. Trends Cell Biol. 11, 294-297, (2001)

[40] Yamano T, Murata S, Shimbara N *et al.* Two distinct pathways mediated by PA28 and hsp90 in major histocompatibility complex class I antigen processing. *J Exp Med.* 196, 185-196 (2002).

[41] Yamano T, Mizukami S, Murata S, Chiba T, Tanaka K, Udono H. Hsp90-mediated assembly of the 26 S proteasome is involved in major histocompatibility complex class I antigen processing. *J Biol Chem.* 283, 28060-28065 (2008).

[42] Wigley WC, Fabunmi RP, Lee MG, Marino CR, Muallem S, DeMartino GN, Thomas PJ. Dynamic association of proteasomal machinery with the centrosome. *J Cell Biol.* 145, 481-490, (1999).

[43] Kamiguchi K, Torigoe T, Fujiwara O, Ohshima S, Hirohashi Y, Sahara H, Hirai I, Kohgo Y, Sato N. Disruption of the association of 73 kDa heat shock cognate protein with transporters associated with antigen processing (TAP) decreases TAP-dependent translocation of antigenic peptides into the endoplasmic reticulum. *Microbiol Immunol.* 52, 94-106, (2008).

[44] Chen D, Androlewicz MJ. Heat shock protein 70 moderately enhances peptide binding and transport by the transporter associated with antigen processing. *Immunol Lett.* 75, 143-148, (2001).

[45] Kropp LE, Garg M, Binder RJ. Ovalbumin-derived precursor peptides are transferred sequentially from gp96 and calreticulin to MHC class I in the endoplasmic reticulum. *J Immunol.* 184, 5619-5627, (2010).

[46] Li Z, Srivastava PK. Tumor rejection antigen gp96/grp94 is an ATPase: implications for protein folding and antigen presentation. *EMBO J.* 12, 3143-3151, (1993).

[47] Serwold T, Gonzalez F, Kim J, Jacob R, Shastri N. ERAAP customizes peptides for MHC class I molecules in the endoplasmic reticulum. *Nature.* 419, 480-483, (2002).

[48] Saric T, Chang SC, Hattori A, York IA, Markant S, Rock KL, Tsujimoto M, Goldberg AL. An IFN-gamma-induced aminopeptidase in the ER, ERAP1, trims precursors to MHC class I-presented peptides. *Nat Immunol.* 3, 1169-1176, (2002).

[49] York IA, Chang SC, Saric T, Keys JA, Favreau JM, Goldberg AL, Rock KL. The ER aminopeptidase ERAP1 enhances or limits antigen presentation by trimming epitopes to 8-9 residues. *Nat Immunol.* 3, 1177-1184, (2002).

[50] Ménoret A, Peng P, Srivastava PK. Association of peptides with heat shock protein gp96 occurs in vivo and not after cell lysis. *Biochem Biophys Res Commun.* 262, 813-818, (1999).
[51] De Maio, A. Extracellular heat shock proteins, cellular export vesicles, and the Stress Observation System: a form of communication during injury, infection, and cell damage. It is never known how far a controversial finding will go! *Cell Stress Chaperones.* 16, 235-249, (2011).
[52] Pawaria S, Messmer MN, Zhou YJ, Binder RJ. A role for the heat shock protein-CD91 axis in the initiation of immune responses to tumors. *Immunol Res.* 50, 255-260, (2011).
[53] Lotze MT, Zeh HJ, Rubartelli A, Sparvero LJ, Amoscato AA, Washburn NR, Devera ME, Liang X, Tör M, Billiar T. The grateful dead: damage-associated molecular pattern molecules and reduction/oxidation regulate immunity. *Immunol Rev.* 220, 60-81, (2007).
[54] Binder RJ. Hsp receptors: the cases of identity and mistaken identity. *Curr Opin Mol Ther.* 11, 62-71, (2009).
[55] Binder RJ, Han DK, Srivastava PK. CD91: a receptor for heat shock protein gp96. *Nat Immunol.* 1, 151-155 (2000).
[56] Basu S, Binder RJ, Ramalingam T, Srivastava PK. CD91 is a common receptor for heat shock proteins gp96, hsp90, hsp70, and calreticulin. *Immunity.* 14, 303-313 (2001).
[57] Binder RJ, Srivastava PK. Essential role of CD91 in re-presentation of gp96-chaperoned peptides. *Proc. Natl. Acad. Sci. USA.* 101, 6128-6133 (2004).
[58] Pawaria S, Binder RJ. CD91-dependent programming of T-helper cell responses following heat shock protein immunization. *Nat Commun.* 2, 521, (2011). doi: 10.1038/ncomms1524.
[59] den Haan, J.M., Lehar, S.M., & Bevan, M.J. CD8(+) but not CD8(-) dendritic cells cross-prime cytotoxic T cells in vivo. *J Exp Med.* 192, 1685-1696, (2000).
[60] Bedoui, S., Whitney, P.G., Waithman, J., Eidsmo, L., Wakim, L., Caminschi, I., Allan, R.S., Wojtasiak, M., Shortman, K., Carbone, F.R., Brooks, A.G., & Heath, W.R. Cross-presentation of viral and self antigens by skin-derived CD103+ dendritic cells. *Nat Immunol.* 10, 488-495, (2009)
[61] Ichiyanagi T, Imai T, Kajiwara C, Mizukami S, Nakai A, Nakayama T, Udono H. Essential role of endogenous heat shock protein 90 of dendritic cells in antigen cross-presentation. *J Immunol.* 185, 2693-2700, (2010).
[62] Imai T, Kato Y, Kajiwara C, Mizukami S, Ishige I, Ichiyanagi T, Hikida M, Wang JY, Udono H. Heat shock protein 90 (HSP90) contributes to cytosolic translocation of extracellular antigen for cross-presentation by dendritic cells. *Proc Natl Acad Sci U S A.* 108, 16363-16368 (2011).
[63] Oura J, Tamura Y, Kamiguchi K, Kutomi G, Sahara H, Torigoe T, Himi T, Sato N. Extracellular heat shock protein 90 plays a role in translocating chaperoned antigen from endosome to proteasome for generating antigenic peptide to be cross-presented by dendritic cells. *Int Immunol.* 23, 223-237, (2011).
[64] Doody AD, Kovalchin JT, Mihalyo MA, Hagymasi AT, Drake CG, Adler AJ. Glycoprotein 96 Can Chaperone Both MHC Class I- and Class II-Restricted Epitopes for In Vivo Presentation, but Selectively Primes CD8(+) T Cell Effector Function. *J Immunol.* 172, 6087-6092 (2004).

[65] SenGupta D, Norris PJ, Suscovich TJ *et al.* Heat shock protein-mediated cross-presentation of exogenous HIV antigen on HLA class I and class II. *J Immunol.* 173, 1987-1993 (2004).

[66] Tobian AA, Canaday DH, Harding CV. Bacterial heat shock proteins enhance class II MHC antigen processing and presentation of chaperoned peptides to CD4+ T cells. *J Immunol.* 173, 5130-5137 (2004).

[67] Matsutake T, Sawamura T, Srivastava PK. High efficiency CD91- and LOX-1-mediated re-presentation of gp96-chaperoned peptides by MHC II molecules. *Cancer Immun.* 10, 7, (2010).

[68] Li M, Davey GM, Sutherland RM *et al.* Cell-associated ovalbumin is cross-presented much more efficiently than soluble ovalbumin in vivo. *J Immunol.* 166, 6099-6103 (2001).

[69] Binder RJ, Srivastava PK. Peptides chaperoned by heat-shock proteins are a necessary and sufficient source of antigen in the cross-priming of CD8+ T cells. *Nat Immunol.* 6, 593-599 (2005).

[70] Blachere NE, Darnell RB, Albert ML. Apoptotic cells deliver processed antigen to dendritic cells for cross-presentation. *PLoS Biol.* 3: e185 (2005).

[71] Serna, A., Ramirez, M.C., Soukhanova, A., & Sigal, L.J. Cutting edge: efficient MHC class I cross-presentation during early vaccinia infection requires the transfer of proteasomal intermediates between antigen donor and presenting cells. *J. Immunol.* 171, 5668-5672, (2003).

[72] Basta S, Stoessel R, Basler M, van den Broek M, Groettrup M. Cross-presentation of the long-lived lymphocytic choriomeningitis virus nucleoprotein does not require neosynthesis and is enhanced via heat shock proteins. *J Immunol.* 175, 796-805, (2005).

[73] Norbury CC, Basta S, Donohue KB, Tscharke DC, Princiotta MF, Berglund P, Gibbs J, Bennink JR, Yewdell JW. CD8+ T cell cross-priming via transfer of proteasome substrates. *Science.* 304, 1318-1321, (2004).

[74] Basu S, Binder RJ, Suto R, Anderson KM, Srivastava PK. Necrotic but not apoptotic cell death releases heat shock proteins, which deliver a partial maturation signal to dendritic cells and activate the NF-kappa B pathway. *Int Immunol.* 12, 1539-1546 (2000).

[75] Singh-Jasuja H, Scherer HU, Hilf N *et al.* The heat shock protein gp96 induces maturation of dendritic cells and down-regulation of its receptor. *Eur J Immunol.* 30, 2211-2215, (2000).

[76] Panjwani NN, Popova L, Srivastava PK. Heat shock proteins gp96 and hsp70 activate the release of nitric oxide by APCs. *J Immunol.* 168, 2997-3003, (2002).

[77] Lehner T, Bergmeier LA, Wang Y *et al.* Heat shock proteins generate beta-chemokines which function as innate adjuvants enhancing adaptive immunity. *Eur J Immunol.* 30, 594-603, (2000).

[78] Ehrlich, P: *Collected Studies on Immunity*. J. Wiley & Sons, London,England. (1906).

[79] Bashford, E., J. Murray, and M. Haaland: *Resistance and susceptibility to inoculated cancer*. Third Scientific Report on the Investigations of the Imperial Cancer Research Fund. E. Bashford, editor. Taylor & Francis, London, England. 359–397, (1908).

[80] North RJ, Kirstein DP. T-cell-mediated concomitant immunity to syngeneic tumors. I. Activated macrophages as the expressors of nonspecific immunity to unrelated tumors and bacterial parasites. *J Exp Med.* 145, 275-292, (1977).

[81] Yee C, Riddell SR, Greenberg PD. In vivo tracking of tumor-specific T cells. *Curr Opin Immunol.* 13, 141-146, (2001).

[82] Ishii KJ, Suzuki K, Coban C, Takeshita F, Itoh Y, Matoba H, Kohn LD, Klinman DM. Genomic DNA released by dying cells induces the maturation of APCs. *J Immunol.* 167, 2602-2607, (2001).

[83] Martin CA, Carsons SE, Kowalewski R, Bernstein D, Valentino M, and Santiago-Schwarz F. Aberrant Extracellular and Dendritic Cell (DC) Surface Expression of Heat Shock Protein (hsp)70 in the Rheumatoid Joint: Possible Mechanisms of hsp/DC-Mediated Cross-Priming. *J Immunol.* 171, 5736-5742, (2003).

[84] Sedlackova L, Nguyen TT, Zlacka D, Sosna A, Hromadnikova I. Cell surface and relative mRNA expression of heat shock protein 70 in human synovial cells. *Autoimmunity.* 42, 17-24, (2009).

[85] Auger I, Escola JM, Gorvel JP, Roudier J. HLA-DR4 and HLA-DR10 motifs that carry susceptibility to rheumatoid arthritis bind 70-kD heat shock proteins. *Nat. Med. 2:306,* (1996).

[86] Beutler B. Microbe sensing, positive feedback loops, and the pathogenesis of inflammatory diseases. *Immunol Rev.* 227, 248-263, (2009).

[87] Yang Y, Liu B, Dai J, Srivastava PK, Zammit DJ, Lefrançois L, Li Z. Heat shock protein gp96 is a master chaperone for toll-like receptors and is important in the innate function of macrophages. *Immunity.* 26, 215-226, (2007).

[88] Wu S, Hong F, Gewirth D, Guo B, Liu B, Li Z. The Molecular Chaperone gp96/GRP94 Interacts With Toll-Like Receptors And Integrins Via Its C-Terminal Hydrophobic Domain. *J Biol Chem.* Jan 5. 2012 [Epub ahead of print].

[89] Staron M, Yang Y, Liu B, Li J, Shen Y, Zúñiga-Pflücker JC, Aguila HL, Goldschneider I, Li Z. gp96, an endoplasmic reticulum master chaperone for integrins and Toll-like receptors, selectively regulates early T and B lymphopoiesis. *Blood.* 115, 2380-2390, (2010).

[90] Weekes MP, Antrobus R, Talbot S, Hör S, Simecek N, Smith DL, Bloor S, Randow F, Lehner PJ. Proteomic Plasma Membrane Profiling Reveals an Essential Role for gp96 in the Cell Surface Expression of LDLR Family Members, Including the LDL Receptor and LRP6. *J Proteome Res.* Feb 7. 2012 [Epub ahead of print].

[91] Neckers L, Workman P. Hsp90 molecular chaperone inhibitors: are we there yet? *Clin Cancer Res.* 18, 64-76, (2012).

[92] Kawabe M, Mandic M, Taylor JL, Vasquez CA, Wesa AK, Neckers LM, Storkus WJ. Heat shock protein 90 inhibitor 17-dimethylaminoethylamino-17-demethoxygeldanamycin enhances EphA2+ tumor cell recognition by specific CD8+ T cells. *Cancer Res.* 69, 6995-7003, (2009).

[93] Yedavelli SP, Guo L, Daou ME, Srivastava PK, Mittelman A, Tiwari RK. Preventive and therapeutic effect of tumor derived heat shock protein, gp96, in an experimental prostate cancer model. *Int J Mol Med.* 4, 243-248, (1999).

[94] Mazzaferro V, Coppa J, Carrabba MG *et al.* Vaccination with autologous tumor derived heat shock protein peptide complex Gp-96 (HSPPC-96) following curative resection of colorectal liver metastases. *Clin. Cancer Res.*, 9, 3235-3245, (2003).

[95] Fong L, Hou Y, Rivas A *et al.* Altered peptide ligand vaccination with Flt3 ligand expanded dendritic cells for tumor immunotherapy. *Proc Natl Acad Sci U S A.* 98, 8809-8814 (2001).

[96] Banchereau J, Palucka AK, Dhodapkar M *et al.* Immune and clinical responses in patients with metastatic melanoma to CD34(+) progenitor-derived dendritic cell vaccine. *Cancer Res.* 61, 6451-6458, (2001).

[97] Vonderheide RH, Domchek SM, Schultze JL *et al.* Vaccination of cancer patients against telomerase induces functional antitumor CD8+ T lymphocytes. *Clin Cancer Res.* 10, 828-839 (2004).

[98] Thurner B, Haendle I, Röder C *et al.* Vaccination with mage-3A1 peptide-pulsed mature, monocyte-derived dendritic cells expands specific cytotoxic T cells and induces regression of some metastases in advanced stage IV melanoma. *J Exp Med.* 190, 1669-1678, (1999).

[99] Lee P, Wang F, Kuniyoshi J *et al.* Effects of interleukin-12 on the immune response to a multipeptide vaccine for resected metastatic melanoma. *J Clin Oncol.* 19, 3836-3847 (2001).

[100] Rosenberg SA, Sherry RM, Morton KE *et al.* Tumor progression can occur despite the induction of very high levels of self/tumor antigen-specific CD8+ T cells in patients with melanoma. *J Immunol.* 175, 6169-6176, (2005).

[101] Janetzki S, Palla D, Rosenhauer V, Lochs H, Lewis JJ, Srivastava PK. Immunization of cancer patients with autologous cancer-derived heat shock protein gp96 : A pilot study. *Intl. Journal of Cancer* 88, 232-238, (2000).

[102] Amato RJ, Murray L, Wood L, Savary C. Tomasovic S, Reitsma D. Active specific immunotherapy in patients with renal cell carcinoma (RCC) using autologous tumor derived heat shock protein-peptide comples-96 (HSPP-96) vaccine. Presented at: *ASCO Meeting*, 2000.

[103] Amato RJ, Murray L, Wood L *et al.* Active specific immunotherapy in patients with renal cell carcinoma (RCC) using autologous tumor derived heat shock protein-peptide comples-96 (HSPP-96) vaccine. Presented at: *ASCO Meeting*, 1999.

[104] Eton O, East MJ, Ross M *et al.* Autologous tumor-derived heat shock protein-peptide comples-96 (HSPP-96) in patients (PTS) with metastatic melanoma. *Proc Am Assoc Cancer Res.* 41, 543 (2000).

[105] Rivoltini L, Castelli C, Carrabba M *et al.* Human Tumor-Derived Heat Shock Protein 96 Mediates In Vitro Activation and In Vivo Expansion of Melanoma and Colon Carcinoma-Specific T Cells. *J. Immunology* 171, 3467-3474 (2003).

[106] Li Z, Qiao Y, Liu B *et al.* Combination of imatinib mesylate with autologous leukocyte-derived heat shock protein and chronic myelogenous leukemia. *Clin Cancer Res.* 11, 4460-4468 (2005).

[107] See AP, Pradilla G, Yang I, Han S, Parsa AT, Lim M. Heat shock protein-peptide complex in the treatment of glioblastoma. *Expert Rev Vaccines.* 10, 721-731, (2011).

[108] Wood C, Srivastava P, Bukowski R, Lacombe L, Gorelov AI, Gorelov S, Mulders P, Zielinski H, Hoos A, Teofilovici F, Isakov L, Flanigan R, Figlin R, Gupta R, Escudier B; C-100-12 RCC Study Group: An adjuvant autologous therapeutic vaccine (HSPPC-96; vitespen) versus observation alone for patients at high risk of recurrence after nephrectomy for renal cell carcinoma: a multicentre, open-label, randomised phase III trial. *Lancet.* 372, 145-154, (2008).

[109] Testori A, Richards J, Whitman E, Mann GB, Lutzky J, Camacho L, Parmiani G, Tosti G, Kirkwood JM, Hoos A, Yuh L, Gupta R, Srivastava PK; C-100-21 Study Group. Phase III comparison of vitespen, an autologous tumor-derived heat shock protein gp96

peptide complex vaccine, with physician's choice of treatment for stage IV melanoma: the C-100-21 Study Group. *J Clin Oncol.* 26, 955-962, (2008).

[110] Mo A, Musselli C, Chen H, Pappas J, Leclair K, Liu A, Chicz RM, Truneh A, Monks S, Levey DL, Srivastava PK. A heat shock protein based polyvalent vaccine targeting HSV-2: CD4(+) and CD8(+) cellular immunity and protective efficacy. *Vaccine.* 29, 8530-8541, (2011).

[111] Wald A, Koelle DM, Fife K, Warren T, Leclair K, Chicz RM, Monks S, Levey DL, Musselli C, Srivastava PK. Safety and immunogenicity of long HSV-2 peptides complexed with rhHsc70 in HSV-2 seropositive persons. *Vaccine.* 29, 8520-8529, (2011).

[112] Wang S, Qiu L, Liu G, Li Y, Zhang X, Jin W, Gao GF, Kong X, Meng S. Heat shock protein gp96 enhances humoral and T cell responses, decreases Treg frequency and potentiates the anti-HBV activity in BALB/c and transgenic mice. *Vaccine.* 29, 6342-6351, (2011).

[113] Stebbing J, Gazzard B, Kim L, Portsmouth S, Wildfire A, Teo I, Nelson M, Bower M, Gotch F, Shaunak S, Srivastava P, Patterson S. The heat-shock protein receptor CD91 is up-regulated in monocytes of HIV-1-infected "true" long-term nonprogressors. *Blood.* 101, 4000-4004, (2003).

[114] Stebbing J, Gazzard B, Portsmouth S, Gotch F, Kim L, Bower M, Mandalia S, Binder R, Srivastava P, Patterson S. Disease-associated dendritic cells respond to disease-specific antigens through the common heat shock protein receptor. *Blood.* 102, 1806-1814, (2003).

[115] Kebba A, Stebbing J, Rowland S, Ingram R, Agaba J, Patterson S, Kaleebu P, Imami N, Gotch F. Expression of the common heat-shock protein receptor CD91 is increased on monocytes of exposed yet HIV-1-seronegative subjects. *J Leukoc Biol.* 78, 37-42, (2005).

[116] Chandawarkar RY, Wagh MS, Srivastava PK. The dual nature of specific immunological activity of tumor-derived gp96 preparations. *J Exp Med.* 189, 1437-1442 (1999).

[117] Chandawarkar RY, Wagh MS, Kovalchin JT, Srivastava P. Immune modulation with high-dose heat-shock protein gp96: therapy of murine autoimmune diabetes and encephalomyelitis. *Int Immunol.* 16, 615-624, (2004).

[118] Kovalchin JT, Mendonca C, Wagh MS, Wang R, Chandawarkar RY. In vivo treatment of mice with heat shock protein, gp 96, improves survival of skin grafts with minor and major antigenic disparity. *Transpl Immunol.* 15, 179-185, (2006).

[119] Liu Z, Li X, Qiu L, Zhang X, Chen L, Cao S, Wang F, Meng S. Treg suppress CTL responses upon immunization with HSP gp96. *Eur J Immunol.* 39, 3110-3120, (2009).

[120] Ogden CA, deCathelineau A, Hoffmann PR, Bratton D, Ghebrehiwet B, Fadok VA, Henson PM. C1q and mannose binding lectin engagement of cell surface calreticulin and CD91 initiates macropinocytosis and uptake of apoptotic cells. *J Exp Med.* 194, 781-795, (2001).

[121] Banerjee PP, Vinay DS, Mathew A, Raje M, Parekh V, Prasad DV, Kumar A, Mitra D, Mishra GC. Evidence that glycoprotein 96 (B2), a stress protein, functions as a Th2-specific costimulatory molecule. *J Immunol.* 169, 3507-3518, (2002).

[122] Delneste Y, Magistrelli G, Gauchat J, Haeuw J, Aubry J, Nakamura K, Kawakami-Honda N, Goetsch L, Sawamura T, Bonnefoy J, Jeannin P. Involvement of LOX-1 in dendritic cell-mediated antigen cross-presentation. *Immunity.* 17, 353-362, (2002).

[123] Vandivier RW, Ogden CA, Fadok VA, Hoffmann PR, Brown KK, Botto M, Walport MJ, Fisher JH, Henson PM, Greene KE. Role of surfactant proteins A, D, and C1q in the clearance of apoptotic cells in vivo and in vitro: calreticulin and CD91 as a common collectin receptor complex. *J Immunol.* 169, 3978-3986, (2002).

[124] Gardai SJ, Xiao YQ, Dickinson M, Nick JA, Voelker DR, Greene KE, Henson PM. By binding SIRPalpha or calreticulin/CD91, lung collectins act as dual function surveillance molecules to suppress or enhance inflammation. *Cell.* 115, 13-23, (2003).

[125] Thrombospondin signaling through the calreticulin/LDL receptor-related protein co-complex stimulates random and directed cell migration. Orr AW, Elzie CA, Kucik DF, Murphy-Ullrich JE. *J Cell Sci.* 2003 Jul 15;116(Pt 14):2917-27.

[126] Orr AW, Pedraza CE, Pallero MA, Elzie CA, Goicoechea S, Strickland DK, Murphy-Ullrich JE. Low density lipoprotein receptor-related protein is a calreticulin coreceptor that signals focal adhesion disassembly. *J Cell Biol.* 161, 1179-1189. (2003). Erratum in: *J Cell Biol.* 2003 Aug 4;162(3):521.

[127] Tobian AA, Canaday DH, Boom WH, Harding CV. Bacterial heat shock proteins promote CD91-dependent class I MHC cross-presentation of chaperoned peptide to CD8+ T cells by cytosolic mechanisms in dendritic cells versus vacuolar mechanisms in macrophages. *J Immunol.* 172, 5277-5286, (2004).

[128] Langelaar MF, Hope JC, Rutten VP, Noordhuizen JP, van Eden W, Koets AP. Mycobacterium avium ssp. paratuberculosis recombinant heat shock protein 70 interaction with different bovine antigen-presenting cells. *Scand J Immunol.* 61, 242-250, (2005).

[129] Tobian AA, Harding CV, Canaday DH. Mycobacterium tuberculosis heat shock fusion protein enhances class I MHC cross-processing and -presentation by B lymphocytes. *J Immunol.* 174, 5209-5214, (2005).

[130] Gardai SJ, McPhillips KA, Frasch SC, Janssen WJ, Starefeldt A, Murphy-Ullrich JE, Bratton DL, Oldenborg PA, Michalak M, Henson PM. Cell-surface calreticulin initiates clearance of viable or apoptotic cells through trans-activation of LRP on the phagocyte. *Cell.* 123, 321-334, (2005).

[131] Ling S, Pi X, Holoshitz J. The rheumatoid arthritis shared epitope triggers innate immune signaling via cell surface calreticulin. *J Immunol.* 179, 6359-6367, (2007).

[132] Shelburne CE, Coopamah MD, Sweier DG, An FY, Lopatin DE. HtpG, the Porphyromonas gingivalis HSP-90 homologue, induces the chemokine CXCL8 in human monocytic and microvascular vein endothelial cells. *Cell Microbiol.* 9, 1611-1619, (2007).

[133] Cheng CF, Fan J, Fedesco M, Guan S, Li Y, Bandyopadhyay B, Bright AM, Yerushalmi D, Liang M, Chen M, Han YP, Woodley DT, Li W. Transforming growth factor alpha (TGFalpha)-stimulated secretion of HSP90alpha: using the receptor LRP-1/CD91 to promote human skin cell migration against a TGFbeta-rich environment during wound healing. *Mol Cell Biol.* 28, 3344-3358, (2008).

[134] Robert J, Ramanayake T, Maniero GD, Morales H, Chida AS. Phylogenetic conservation of glycoprotein 96 ability to interact with CD91 and facilitate antigen cross-presentation. *J Immunol.* 180, 3176-3182, (2008).

[135] De Filippo A, Binder RJ, Camisaschi C, Beretta V, Arienti F, Villa A, Della Mina P, Parmiani G, Rivoltini L, Castelli C. Human plasmacytoid dendritic cells interact with gp96 via CD91 and regulate inflammatory responses. *J Immunol.* 181, 6525-6535, (2008).

[136] Woodley DT, Fan J, Cheng CF, Li Y, Chen M, Bu G, Li W. Participation of the lipoprotein receptor LRP1 in hypoxia-HSP90alpha autocrine signaling to promote keratinocyte migration. *J Cell Sci.* 2009 122, 1495-1498, (2009).

[137] Fischer N, Haug M, Kwok WW, Kalbacher H, Wernet D, Dannecker GE, Holzer U. Involvement of CD91 and scavenger receptors in Hsp70-facilitated activation of human antigen-specific CD4+ memory T cells. *Eur J Immunol.* 40, 986-997, (2010).

[138] Chen JS, Hsu YM, Chen CC, Chen LL, Lee CC, Huang TS. Secreted heat shock protein 90alpha induces colorectal cancer cell invasion through CD91/LRP-1 and NF-kappaB-mediated integrin alphaV expression. *J Biol Chem.* 285, 25458-25466, (2010).

[139] Gopal U, Bohonowych JE, Lema-Tome C, Liu A, Garrett-Mayer E, Wang B, Isaacs JS. A novel extracellular Hsp90 mediated co-receptor function for LRP1 regulates EphA2 dependent glioblastoma cell invasion. *PLoS One.* 6, e17649, (2011).

[140] Tischer S, Basila M, Maecker-Kolhoff B, Immenschuh S, Oelke M, Blasczyk R, Eiz-Vesper B. Heat shock protein 70/peptide complexes: potent mediators for the generation of antiviral T cells particularly with regard to low precursor frequencies. *J Transl Med.* 9, 175, (2011).

In: Heat Shock Proteins
Editor: Saad Usmani

ISBN: 978-1-62417-571-8

Chapter II

Heat Shock Proteins as Molecular Targets for Anticancer Therapy: Approaches, Agents, and Trends

Alexander E. Kabakov*[*] *and Vladimir A. Kudryavtsev
Department of Radiation Biochemistry, Medical Radiology Research Center, Obninsk, Russia

Abstract

Inducible forms of major heat shock proteins (Hsp90, Hsp70 and Hsp27) can be implicated in carcinogenesis as onco-promoters and many human malignant tumors exhibit the enhanced Hsp expression. Besides, the *in vivo* induction and accumulation of Hsps in tumors may be stimulated by pathophysiological states accompanying malignant growth (e.g. hypoxia, acidosis, inflammation etc) or some kinds of anticancer therapy (e.g. hyperthermia, certain drugs). Importantly, the increased Hsp level in human malignancies is associated with their resistance to therapeutics and poor outcome for patients. Numerous facts indicate that the enhanced Hsp expression in solid tumors promotes their unlimited growth, invasion and metastasis formation. Furthermore, excess Hsps can protect cancer cells from apoptosis, senescence and mitotic catastrophe thus contributing to their resistance to chemotherapy and radiotherapy. On the contrary, artificial inhibition of expression or functional activities of Hsps in cancer cells impairs their proliferative and invasive potential and sensitizes them to hyperthermia, chemotherapy and radiotherapy. All this allows to consider inducible Hsps as promising molecular targets and motivates efforts on creation of clinically applicable inhibitors of the Hsp expression/activity in patients' tumors. The present chapter reviews various approaches to targeting Hsps and Hsp-involving pathways in cancer cells to achieve direct beneficial effects or the synergism with conventional therapeutics. Encouraging advances in development of Hsp-inhibiting antitumor agents on the basis of natural or synthetic compounds and peptide or nucleotide sequences are described in the chapter, while some problems and perspectives in this field are also discussed.

---

[*] E-mail: aekabakov@hotmail.com.

Introduction

In mammalian cells, the so called ‘heat stress response’ is triggered by a heat shock transcription factor 1 (HSF1) whose activation results from the heat-induced denaturation of cellular proteins [1, 2]. The activated (trimeric) HSF1 binds to the specific nucleotide sequence, called heat shock element (HSE), in the promoter region of *heat shock*-genes, thereby initiating their transcription (Figure 1). These genes encode a set of cytoprotective ‘heat shock proteins’ (Hsps) with different molecular mass: Hsp27, Hsp40, Hsp70, Hsp90 and others. When these inducible Hsps accumulate in the cell, they can contribute to the post-stress recovery of cellular homeostasis and confer cell tolerance to the stronger, otherwise lethal heating [1-4].

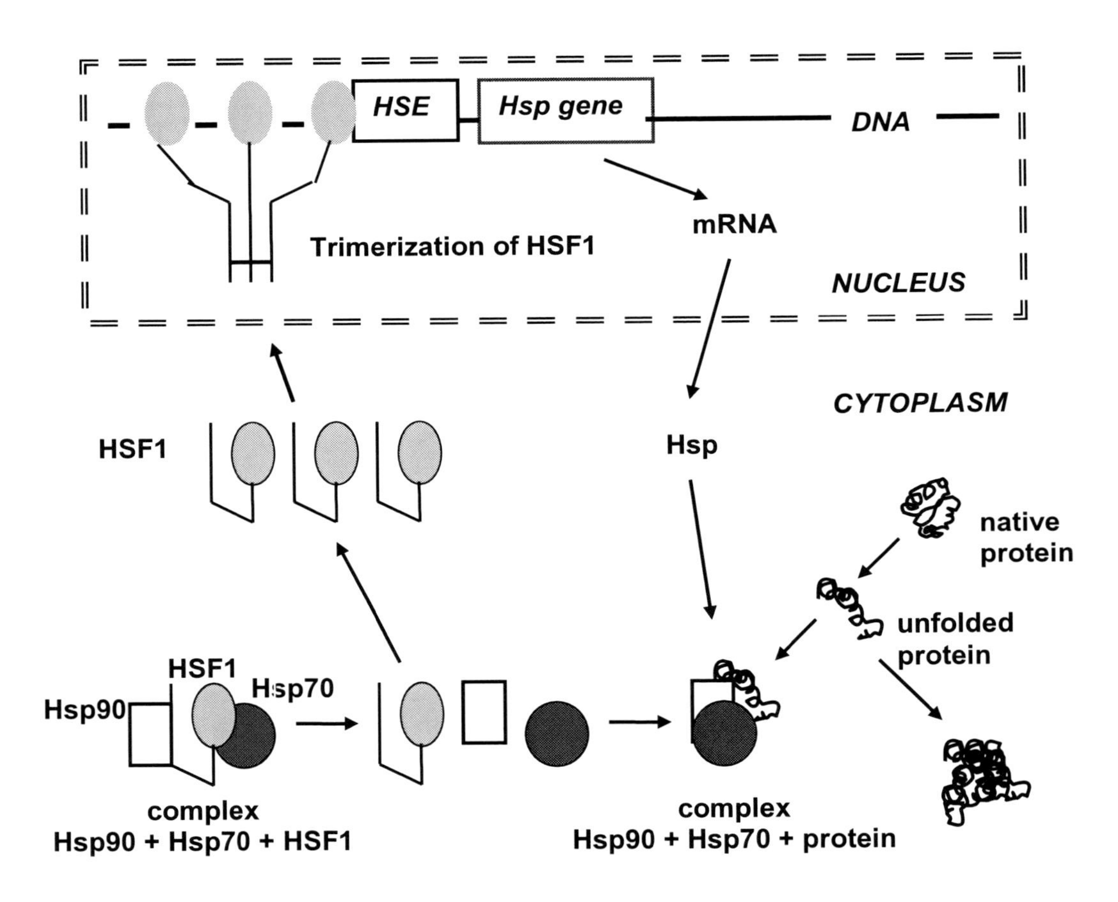

Figure 1. Scheme illustrating Hsp90-, Hsp70-dependent regulation of HSF1-mediated heat stress transcriptional response in mammalian cells (see details in text and also refs [1, 2]).

Besides heat shock, many other stressful or pathological states (e.g. oxidative stress, acidosis, hypoxia, ischemia/reperfusion, inflammation, etc) leads to the HSF1-mediated Hsp induction followed by acquisition of a transient cell tolerance to either stress. Such ‘cross tolerance’ is mainly based on a capability of inducible Hsps to protect from the proteotoxic (protein-damaging) effects of stress because all major Hsps are molecular chaperones that stabilize the native conformations of protein molecules and preserve nascent polypeptides from misfolding [3, 4]. In the unstressed cell, the constitutively expressed Hsp70 and Hsp90 assist protein maturation and transport, proteolysis, and intracellular signal transduction [5, 6]; moreover, both these Hsps bind and inhibit HSF1 thus preventing its activation under normal conditions [1, 2]. As soon as damaged proteins appear in the stressed cell, they recruit Hsp70 and Hsp90 from their inhibitory complex with HSF1 so that the latter is activated via its trimerization and triggers expression of all inducible Hsps (Figure 1).

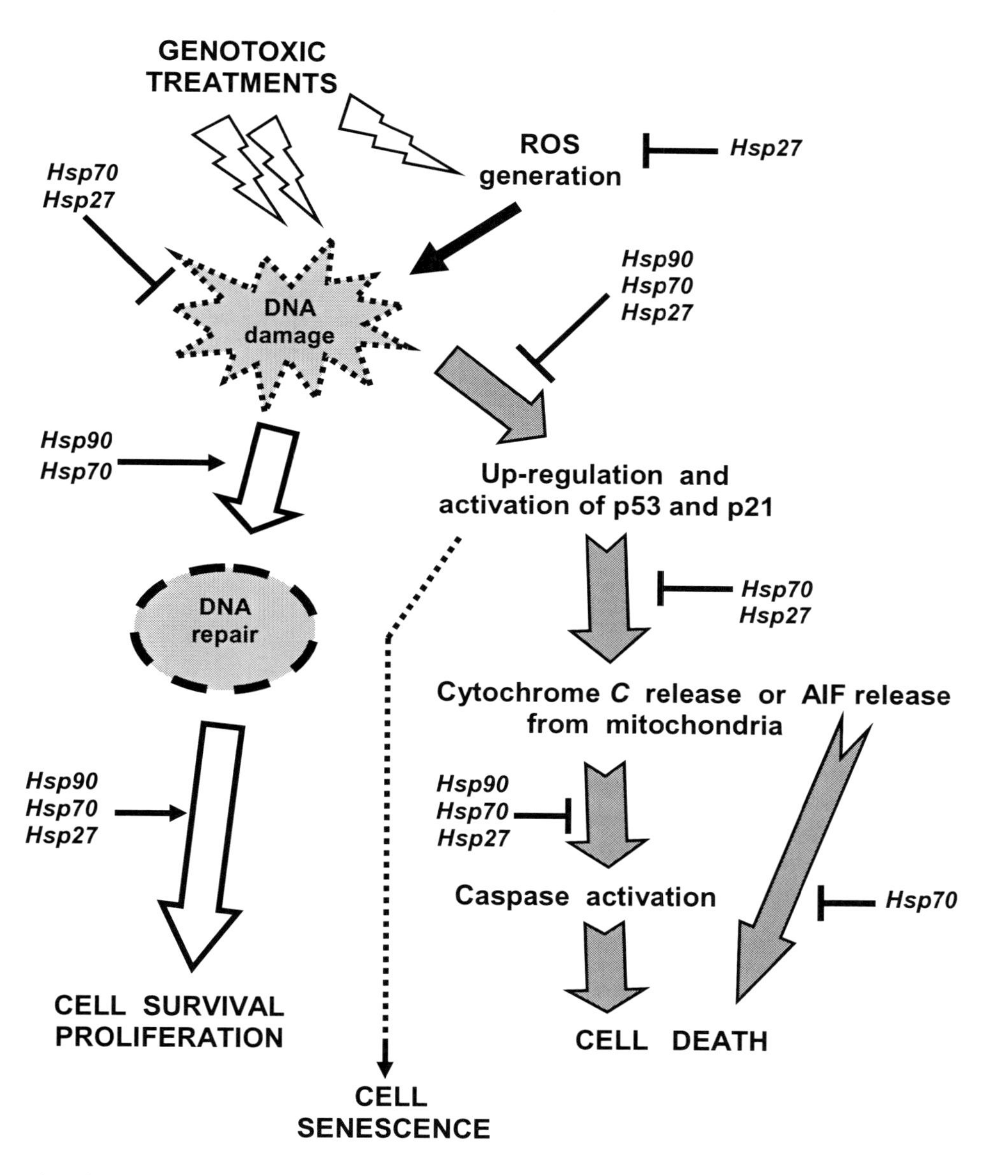

Figure 2. Schematic illustration showing key points for the cancer cell-protecting action of intratumor Hsp90, Hsp70 and Hsp27 against genotoxic (therapeutic) treatments. The stimulating effects of either Hsp are designated by black arrows, whereas the inhibitory effects are designated by black perpendiculars.

In the Hsp-enriched (tolerant) cell undergoing any proteotoxic stress, excess Hsps minimize protein aggregation and then promote refolding or degradation of stress-denatured proteins [1-4]. Upon fulfillment of their restoring functions, liberated Hsp90 and Hsp70 again bind HSF1 and turn off the transcription of *hsp*-genes (Figure 1). Although Hsp27 (referred to as Hsp25 in rodent cells) is not implicated in the HSF1 regulation, this “small” chaperone also protects cellular proteins from aggregation and particularly protects the F-actin cytoskeleton [3, 4]. Moreover, Hsp27 participates in redox control in the cell and can attenuate oxidative stress [7]. Besides their capability to attenuate the stress-associated proteotoxicity, both Hsp90 and Hsp70 and Hsp27 are involved in intracellular signaling to block apoptotic pathways in stressed cells thereby contributing to their survival (see refs [*8-10*] and Figure 2). Additionally, the increased Hsp level may preserve the stress-damaged cell from necrotic death, premature senescence and mitotic catastrophe (reviewed in [11, 12]).

All these cytoprotective activities of inducible Hsps are obviously undesirable and dangerous when they are manifested in cancer cells. In fact, the up-regulated expression of Hsp90, Hsp70 or Hsp27 is found in many cases of human malignancies and often associated with weak tumor response to therapeutics and poor outcome for patients [12, 13]. It was established on numerous models that Hsps overexpressed in cancer cells enhance their proliferative, invasive and angiogenic capacities, and protect them from the injurious factors of tumor microenvironment such as hypoxia and inflammation or immune attacks with TNF and Fas stimulation. Furthermore, the Hsp up-regulation in tumors can confer their resistance to chemotherapy, radiotherapy and hyperthermia. Consequently, HSF1 and inducible Hsps are regarded as potential targets for antitumor therapy. Here we review this problem at the molecular level and describe various approaches to targeting Hsps for fight against cancer.

Hsps Can Protect Cancer Cells from Apoptosis

It is generally accepted that Hsp90, Hsp70 and Hsp27 play a role of suppressors of apoptosis in cancer cells (see for a review [8-12]). In fact, all major apoptotic pathways in the cell can be blocked by these Hsps at different points of the apoptotic signal transduction.

Hsp90

Hsp90 is an abundant cytosolic chaperone whose functioning is necessary for the stabilization and activation of many client proteins including growth factor receptors and serine/threonine kinase Akt which trigger a cascade of cytoprotective reactions aimed at inhibition of mitochondria-, caspase-dependent apoptosis [13-15].

The other key component of Hsp90-mediated protection of cancer cells from apoptosis is survivin – an antiapoptotic oncoprotein whose expression and stability require the Hsp90-conducted chaperoning [16]. Besides, Hsp90 inhibits formation of a functional apoptosome by directly interacting with Apoptosis-activating factor 1 (Apaf-1) [17]; this prevents caspase-9 activation followed by the lethal activation of caspase-3 and caspase-dependent DNase (CAD) (Figure 2).

In the pathway of TNF-receptor 1 (TNFR1)-stimulated signaling, Hsp90 stabilizes the Receptor-Interacting Protein (RIP), thus promoting activation of the transcription nuclear factor kappaB (NF-κB) activation that protects cancer cells from TNF-induced apoptosis [18]. Small-molecule inhibitors of the Hsp90 chaperone activity are known which enable to sensitize malignancies to the apoptosis-inducing action of antitumor therapeutics (see below).

Hsp70

In turn, Hsp70 also interacts with Apaf-1 thus blocking apoptosome formation [19, 20]. There is an additional pathway of Hsp70-induced prevention of apoptosome formation: excess Hsp70 inhibits pro-apoptotic changes in conformation of Bax and its translocation to mitochondria, and downstream Bax-provoked events (cytochrome *c* release into the cytosol, activation of caspase-9, caspase-3, etc) [21, 22].

Moreover, overexpressed Hsp70 is able to suppress caspase-independent apoptosis which is executed by apoptosis-inducing factor (AIF) released from mitochondria (Figure 2); it has been suggested that Hsp70 specifically binds to AIF thereby preventing the AIF-induced chromatin condensation and DNA fragmentation [23].

In some cases, the antiapoptotic capacity of Hsp70 appears to be realized via interruption of 'cell death signals' resulting from the c-Jun N-terminal kinase (JNK) activation in stressed or TNF-treated cells [24, 25]; in this mechanism, excess Hsp70 promotes rapid deactivation (dephosphorylation) of activated JNK [26].

Hsp27

Similarly, Hsp27 can act as a blocker of caspase-dependent apoptosis by interfering with apoptosome formation: Hsp27 is able to directly bind cytosolic cytochrome *c* released from mitochondria, thus sequestering cytochrome *c* from its association with Apaf-1 (Figure 2 and [27]).

Moreover, excess Hsp27 appears to prevent the cytochrom c release from mitochondria following apoptotic stimuli [28]. Among other antiapoptotic activities attributed to Hsp27 there are inhibition of the TNF-stimulated apoptosis via direct interaction of phosphorylated Hsp27 with DAXX [29] and blockade of the Fas-mediated apoptotic signals at the stage of recruitment of the adaptor molecules Fas Associated Death Domain (FADD) via regulation of PEA-15 [30].

In addition, Hsp27 may suppress apoptosis thanks to its capability to facilitate proteasomal degradation of IκB, thereby augmenting the functional activity of NF-κB, a powerful pro-survival effector [31]. There are published data suggesting that overexpressed Hsp27 modulates signaling cascades mediated by insulin-like growth factor 1 (IGF-1) [32] and protein kinases Akt [32-35], PKCδ [36, 37] and ERK 1/2 [38, 39], while such modulation of the signal transduction can confer cancer cell protection from apoptosis.

Finally, Hsp27 was shown to modulate expression of eukaryotic translation initiation factor 4E (elF4E) that plays a role in resistance of androgen-dependent tumor cells to apoptosis resulting from androgen ablation or chemotherapy with paclitaxel [40]. Similar elF4E-involving mechanisms appear to act in the Hsp27-mediated resistance of pancreatic cancer cells to gemcitabine [41].

Apparently, all these multiple antiapoptotic activities of Hsp90, Hsp70 and Hsp27 can contribute to both deregulated growth of the Hsp-enriched tumors and their tolerance to cytotoxic therapeutics. Besides, it seems likely that some of these activities of Hsps (particularly, the activities blocking TNF- and Fas-stimulated apoptosis) help malignancies to resist the antitumor immune response or action of immunotherapy. If so, artificial down-regulation of the Hsp expression and/or activities in cancer cells may exert beneficial (tumor-repressing and tumor-sensitizing) effects.

Hsps can Protect Cancer Cells From Death or Senescence Resulting from DNA Damage

Genotoxic effects exerted by radiation or DNA-damaging antitumor drugs (e.g. etoposide, cisplatin, doxorubicin and others) are the most common cause of cancer cell elimination following radiotherapy or chemotherapy. Indeed, if DNA repair is unsuccessful, the affected cells undergo apoptosis or, instead, mitotic catastrophe or premature senescence [42-45].

After treatments with clinically relevant doses of radiation or genotoxic drugs, apoptosis is the major form of cell death in lymphoid malignancies, whereas premature senescence and mitotic catastrophe rather than apoptosis take place in most typical epithelium-derived solid tumors (carcinomas) [45]. The Hsp expression enhanced in many human tumors appears to confer cancer cell resistance to the genotoxic therapeutics: numerous reports allow to suggest that excess Hsps can contribute to attenuation of DNA damage and/or facilitation of DNA repair and also to switching the DNA damage response from cell death or senescence to cell survival and further proliferation (Figure 2 and refs [11, 12, 46]).

Hsp90

Importantly, small-molecule inhibitors of the Hsp90 chaperone function may cause inactivation and/or early proteasomal degradation of these proteins in target cancer cells undergoing genotoxic (therapeutic) exposures (see the next subsections); in this case, the Hsp90 inhibition will lead to unsuccessful DNA repair followed by massive cell death or premature senescence within the treated cancer cell populations (Figure 2). It was shown in various models that Hsp90 inhibitors do increase DNA damage and/or impair DNA repair in irradiated tumor cells of different origin (see Figure 3 and also refs [49, 50]). The recent data of proteomic studies on human cancer HeLa cells treated with an Hsp90 inhibitor, 17DMAG, confirm that the major Hsp90 inhibition-sensitive molecular targets are kinases and components of the DNA damage response [51].

Our own experimental data demonstrate that the Hsp90 inhibition in human breast cancer MCF-7 cells incubated with 17AAG prior to and following radiation exposure strongly slows down formation and decay of the phosphorylated histone gamma H2AX foci at sites of double-strand (ds) breaks in nuclear DNA (Figure 3) that implies a retardation of the post-radiation DNA repair. This delay in repair of the DNA ds breaks was well correlated with radiosensitization of the 17AAG-treated MCF-7 cells (see below).

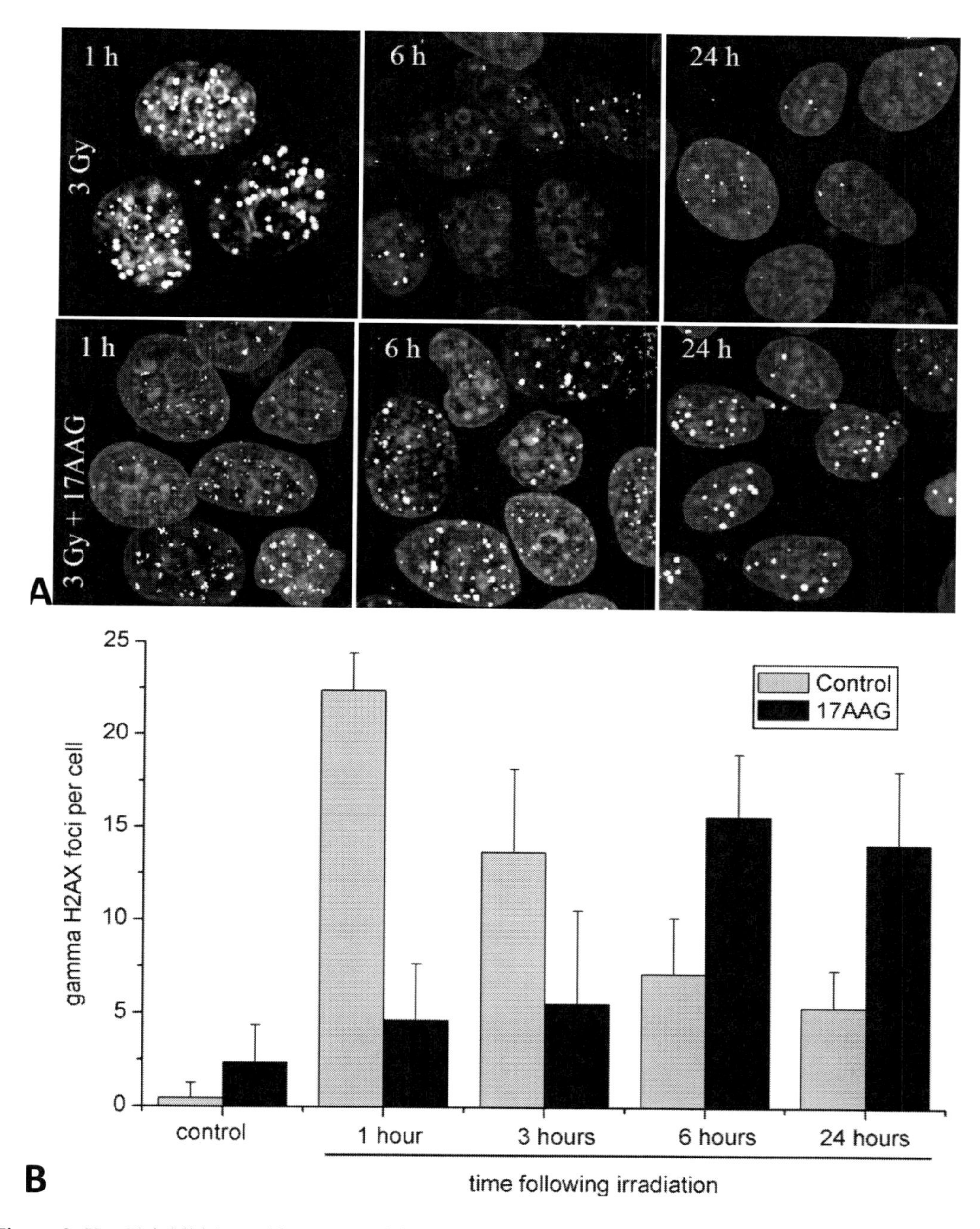

Figure 3. Hsp90 inhibition with 17AAG delays post-radiation repair of dsDNA breaks in human breast cancer MCF-7 cells. (A) – immunofluorescence patterns showing intranuclear foci of phosphorylated histone gamma H2AX (sites of DNA ds breakage/repair) at different time points following 3 Gy γ-irradiation alone (upper panel) or co-treatment 35 nM 17AAG + 3 Gy (lower panel). (B) – Bars showing average amounts ± S.E. of foci per nucleus at different time points following the (co-)treatments; see details in text. (V.A. Kudryavtsev, before unpublished data.)

Among identified client proteins of Hsp90 there are telomerase hTERT, mutated p53, ataxia-telangiectasia mutated (ATM) and cyclin-dependent kinase 4 (CDK4) which are involved in mechanisms of cellular defense against senescence and DNA breakage [11, 13, 15]. Likewise, the Hsp90 chaperone circuit is required for the system of cell cycle checkpoint control that defines the DNA damage response and further fate (death or survival?) of the affected cell (Figure 2 and [11, 13, 15]). Catalytic activities of some DNA break-repairing enzymes such as DNA-dependent protein kinase (DNA-PK), apurinic/apyrimidinic endonuclease and DNA polymerase-beta are also enhanced by Hsp90 [47, 48].

Hsp70

A number of publications indicate that Hsp70 protects cancer cells from genotoxic effects of radiation and chemotherapeutics (see for review [11, 12]). According to some observations, excess Hsp70 can somehow ensure attenuation of DNA damage and/or acceleration of DNA repair. So, it was found that Hsp70 directly interacts with HAP-1 endonuclease, a key enzyme in base-excision repair, and stimulates its DNA-repairing activity [52]. Overexpression of Hsp70 in human lung carcinoma A549 cells did decrease DNA damage resulting from UVC exposure [53]. siRNA-induced Hsp70 depletion in human leukemic cells [54, 55] and human colon carcinoma HCT116 cells [12, 56] inhibited repair of abasic sites in damaged DNA and sensitized the target cells to gamma- or UV-radiation and doxorubicin. Some data suggest that even if Hsp70 does not attenuate DNA damage in the stressed cells, this Hsp may increase the post-stress cell survival by modulating the DNA damage response in favor of DNA repair instead of apoptosis, senescence, and mitotic catastrophe; such Hsp70-mediated mechanism appears to be realized via p53/p21 signaling [12, 57].

While pro-apoptotic activation of Bax by DNA damage may be p53-dependent or p53-independent, Hsp70 is able to interfere with both these reactions. For example, in etoposide-treated HL-60 cells, Hsp70 was shown to inhibit the p53-mediated Bax conformation changes and Bax translocation to mitochondria as well as subsequent execution of the apoptotic pathway (cytochrom *c* release, activation of caspase-9 and -3, etc) [21]. In the case of p53-independent DNA damage response, Hsp70 may prevent the Bax activation/translocation via either inhibition of Bid cleavage by caspase-2 or suppression of JNKJ/p38 kinase activation [12] (similarly to what happens under non-genotoxic stresses [21, 25]).

As it was demonstrated on HCT116 carcinoma cells exposed to gamma-radiation or doxorubicin, down-regulation of Hsp70 leads to destabilization of Hdm2 that, in turn, results in stabilization of p53 and greater accumulation of its downstream target p21; all those events were accompanied by the tumor cell-sensitizing effects [12, 56]. This finding suggests that Hsp70 can modulate p53 pathway in cancer cells thereby preserving them from senescence and cell death (apoptosis or mitotic catastrophe) following radiation or genotoxic drugs (Figure 2). In addition, Hsp70 may protect tumor cells from p53-independent senescence via modulation of extracellular signal-regulated kinase (ERK) cascade at the level of Raf-1 [58] and activation of the cell cycle kinase Cdc2 [56]. Therefore, artificial inhibition of the Hsp70 expression/activity in tumors would help to increase the efficacy of radiotherapy and chemotherapy.

Hsp27

In some cases of DNA-damaging treatments, Hsp27 seems to act by the same way as Hsp70 does (see Figure 2). So, siRNA-induced Hsp27 down-regulation in human colon carcinoma cells sensitized them to doxorubicin and UV by activating the apoptotic or senescence program via the Hdm2 destabilization and p53/p21 pathway [59, 60]. As well as in the case of Hsp70 depletion [12, 56], activation of p53 and accumulation of p21 took place in the Hsp27-depleted carcinoma cells which hence became more sensitive to doxorubicin [59, 60]. Additional similarity in the cytoprotective actions of Hsp27 and Hsp70 is manifested in a capability of Hsp27 to suppress pro-apoptotic activation of JNK/p38 kinases as it was shown for Hsp27-expressing leukemic cells treated with etoposide [61].

However, Hsp27 may protect from genotoxic treatments by mechanisms different from those of Hsp70. For instance, the Hsp27-conferred resistance of L929 and HeLa cells to cisplatin was associated with Hsp27-mediated enhancement of Akt activation and attenuation of inhibition of thioredoxin reductase [34]. Similarly, down-regulation of Hsp27 in cells of head and neck squamous cell carcinomas resulted in the cell radiosensitization through a decrease in the intracellular levels of reduced glutathione and impairment of the irradiation-responsive Akt activation [35].

Many other publications suggest that Hsp27 protects from (pro)oxidants and free radicals by increasing the cytosolic pool of reduced glutathione (see for review [3, 7]). This mechanism may attenuate oxidative damage to DNA; so, up-regulation of Hsp27 in L929 cells diminished the level of oxidative modification of nuclear DNA following TNFalpha-induced oxidative stress [62].

While the injurious effects of ionizing radiation are partly due to generation of the reactive oxygen species (ROS) in irradiated cells, there are several reports that Hsp27 contributes to cellular radioresistance by attenuating the ROS-associated cytotoxicity (Figure 2 and [7, 37, 39, 63]). In Jurkat cancer cells, the radioprotection conferred by overexpressed Hsp27 was due to suppression of PKCδ activation and impaired ROS generation upon radiation exposure [37]. In another work, it was suggested that excess Hsp27 in target cancer cells can augment the intracellular pool of reduced glutathione, a major scavenger of ROS, thus increasing the post-radiation cell survival [63]. Moreover, catalyzing proteasomal degradation of NF-κB, excess Hsp27 may cause NF-κB-mediated up-regulation of Mn2+-dependent superoxide dismutase (MnSOD), a superoxide anion scavenger; such a radioprotective mechanism was found in radioresistant L929 cells overexpressing Hsp27 [64]. Analogous down-regulation (degradation) of IκB and hyperactivation of NF-κB appear to take place in Hsp27-conferred protection of human leukemia cells from TNFalpha and etoposide [31].

Such a capability of Hsps to protect the cancer cell from radiation, genotoxic drugs and ROS characterizes them as one of the determinants of tumor resistance to therapeutic intervention (Figure 2). Consequently, targeted inhibition of Hsps in malignant cells may enhance the efficacy of chemotherapy and/or radiotherapy.

Targeting Hsps can Sensitize Malignancies to Chemotherapy

The modern anticancer chemotherapy is mainly aimed at the elimination of malignant cells via pharmacological induction of apoptosis and mitotic catastrophes. The induction of cancer cell senescence is also the desirable outcome herein. Unfortunately, some fractions of drug-treated tumor cells may survive by quickly selecting for adaptive mutations or reprogramming regulation to involve alternative pathways that confer drug resistance.

Hsp90

Because Hsp90 inhibition can block multiple Hsp90-dependent pathways contributing to the tumor cell progression and survival [13-15], pharmacological inhibitors may assist conventional chemotherapeutics to repress tumors. At present, a number of small-molecule inhibitors of the Hsp90 chaperone function are known which manifest promising antitumor activities in various cancer-related models. Among these Hsp90 inhibitors there are antibiotics and their analogs (e.g. 17AAG and 17DMAG), purine-scaffold derivatives (e.g. BIIB021), pyrazoles (e.g. CCT-018159), peptides (e.g. shepherdin), and others [13-15, 65]. Development and testing of new low-toxic Hsp90 inhibitors with improved therapeutic properties are actively continued [66, 67].

It was shown on cell lines of pediatric acute lymphoblastic leukemia that combining with 100 nM 17AAG yields the 10-1000-fold decrease in IC_{50} of most of the common anticancer drugs including imatinib, cytarabine, carboplatin, cisplatin, toptecan, methotrexate, cyclophosphamide, vincristine, and etoposide [15, 68]. In many other studies, nanomolar concentrations of Hsp90 inhibitors being combined with taxanes [69-71], cisplatin [72], oxaliplatin [73] and carboplatin [*74*], doxorubicin [75, 76], etoposide [77], ZD1839 ("Iressa") [78], imatinib [79, 80], bortezomib (Velcade) [81, 82] and trastuzumab (Herceptin) [83, 84] synergistically enhanced cell cycle arrest and apoptosis in various cancer cell lines and tumor xenografts in nude mice. At the molecular level, this synergism was associated with the Hsp90 inhibition-induced inactivation/degradation of the Hsp90 client proteins such as Akt, ALK, HER2, Bcr-Abl, CD1, CDK4 and CDK6, Raf-1 and EGF receptor which are involved in tumor growth-stimulating and tumor cell-protecting pathways.

Besides, Hsp90 catalyzes P-glycoprotein (*MDR1*-gene product) that works as a membrane pump performing clearance of the cell from entered toxins. This P-glycoprotein overexpressed in cancer cells decreases intracellular drug concentrations, thus conferring the multi-drug resistance of tumors [85]. If so, the Hsp90 inhibitors may suppress the Hsp90-dependent action of this pump and increase the efficacy of chemotherapy [79].

Importantly, Hsp90 in cancer cells binds 17AAG with ~100-fold higher affinity than Hsp90 in normal cells does it [86]. Such dramatic difference suggests that in the patient's organism undergoing combined chemotherapy, the Hsp90 inhibitor will preferentially target malignant cells and the drug-induced cytotoxicity will be enhanced just within the tumor area.

According to data of the US National Institute of Health, more than a dozen of Hsp90 inhibitors are currently being tested in I-III Phase clinical trials as single agents or as co-agents with other anticancer drugs (see website: www.clinicaltrials.gov). The most advanced candidate is 17AAG which passes Phase III clinical trials as a co-agent with bortezomib for the treatment of patients with multiple myeloma.

Unfortunately, results of most of the clinical trials of Hsp90 inhibitors seem more modest than it was expected following numerous pre-clinical tests. One of the serious problems herein is that the Hsp90 inhibitors often act as activators of HSF1 and inducers of Hsps including cytoprotective Hsp70 and Hsp27 [87]. Apparently, the Hsp90 inhibition-provoked Hsp up-regulation in the drug-treated cancer cells renders them more resistant to the drug-induced cytotoxicity. If so, inhibitors of HSF1-mediated Hsp induction should be used in combination with inhibitors of the Hsp90 chaperone function (see Figures 4 and 5, and refs [15, 88, 89]); such an approach will allow to enhance the tumor-sensitizing effects of the Hsp90 inhibition.

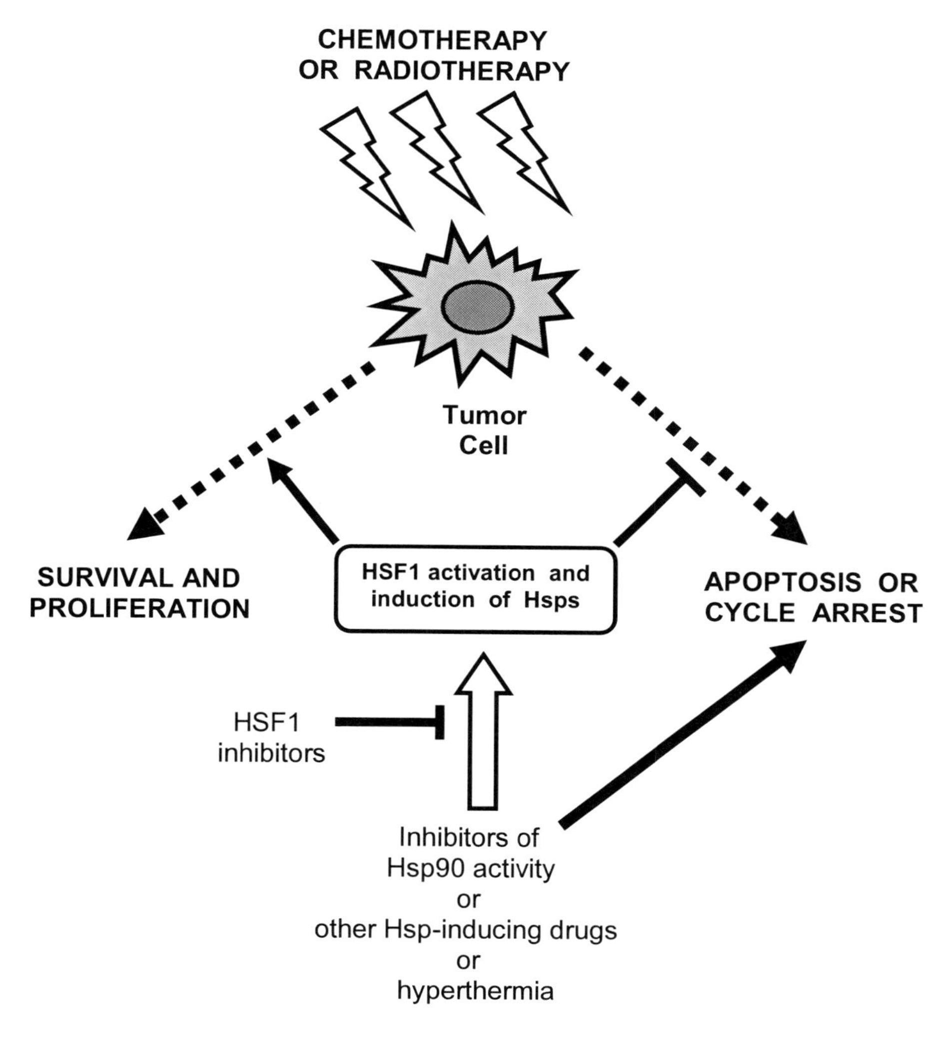

Figure 4. Schematic illustration showing that beneficial (antitumor) effects expected under combining of chemotherapy or radiotherapy with inhibitors of the Hsp90 chaperone activity or hyperthermia may be impaired through the activation of HSF1 and subsequent induction of cytoprotective Hsps. To escape such complications it would be rational to use herein blockers of the HSF1 activation/Hsp induction pathway (see details in text).

In our *in vitro* model studies, we found the significant enhancement of cell death (up to total cell killing) in human leukemic HL-60 cell populations treated with taxol (a cytotoxic anticancer drug) in combination with 17AAG (an Hsp90 inhibitor) and quercetin (a flavonoid inhibiting HSF1 [90]) as compared with the two-drug combination taxol + 17AAG (Figure 5). It was herein revealed by immunoblotting that the addition of quercetin did block the 17AAG-induced up-regulation of Hsp70 and Hsp27 in the treated HL-60 cells (not shown). Although the HSF1-inhibiting (micromolar) concentrations of quercetin are not clinically applicable, our finding gives "proof-of-principle" for the suggested strategy of co-administration of Hsp90 inhibitors with inhibitors of the Hsp induction in order to achieve total elimination of the target cancer cells (see Figures 4 and 5).

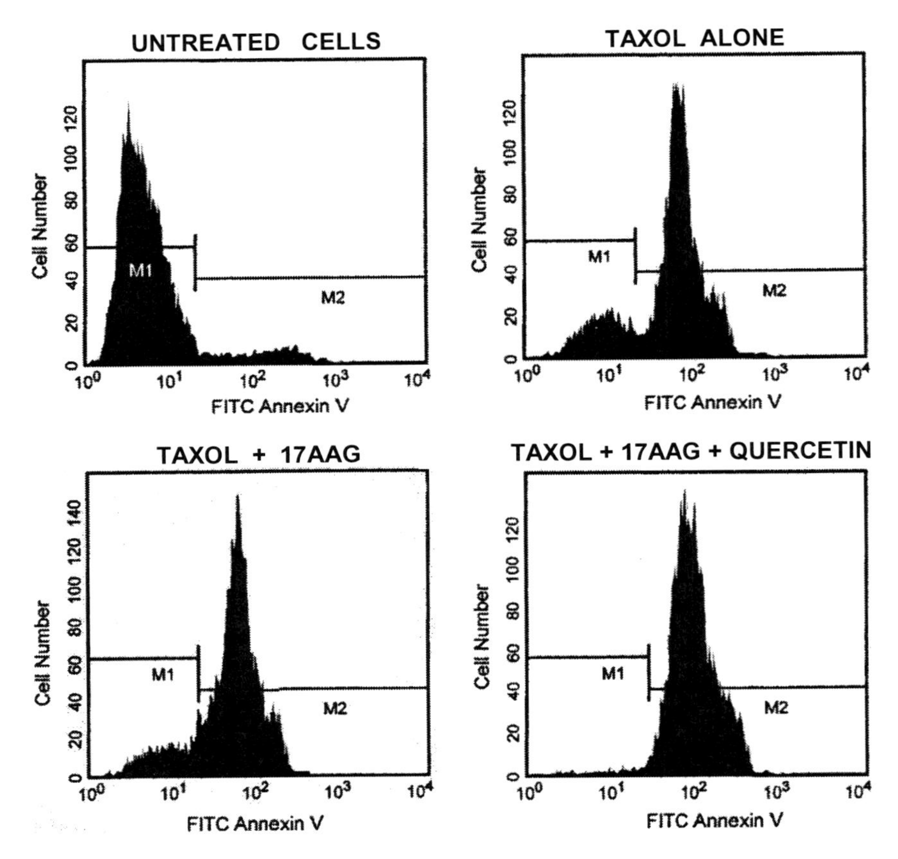

Figure 5. Sensitization of human leukemic cells to taxol by simultaneous inhibiting Hsp90 chaperone activity and HSF1-mediated Hsp induction. HL-60 cells were either untreated or incubated with 10 nM taxol without or with 100 nM 17AAG (an Hsp90 inhibitor) and 30 μM quercetin (an inhibitor of Hsp induction). After 20 h incubation with the drug(s), the cells were stained with FITC-annexin V (a probe to detect dead and dying cells) and analyzed by flow cytometry. Peaks showing the intensity of FITC fluorescence per cell within M1 zone are formed by subpopulations of the viable (FITC-annexin V-negative) cells, whereas peaks within M2 zone present the apoptotic (FITC-annexin V-positive) cells. About 65% of the cells treated with taxol alone were undergoing apoptosis, while many cells (~35%) remained still viable. In the case of co-treatment taxol + 17AAG, a fraction of the viable cells decreased to ~20%. Finally, upon the triple combination: taxol + 17AAG + quercetin, practically total killing of the drug-treated leukemic cells was achieved. (A.E. Kabakov, before unpublished data.)

Similarly, other researchers demonstrated that the Hsp90 inhibition-increased Hsp70 level attenuates apoptosis in 17AAG-treated human leukemia cells, while prevention of such up-regulation of Hsp70 using siRNA or KNK437 can intensify this apoptosis [88]. In another research group, a novel agent NZ28 was developed which can inhibit (at the level of translation) the Hsp70 induction/accumulation in MM.1S myeloma cells and PC3 or DU145 prostate tumor cells treated with Hsp90 inhibitors [89]. McCollum et al [91] found that the Hsp90 inhibition-induced up-regulation of Hsp27 in 17AAG-treated human tumor cells (HeLa, PC3 lines) increases their resistance to 17AAG and this can be abolished with Hsp27 siRNA. All these data provide "proof-of-principle" for an approach with artificial preventing the Hsp induction in Hsp90 inhibitor-treated malignancies (see Figure 4). Hence there is the acute need in development of the clinically available inhibitors of Hsp induction.

It should herein be added that besides the Hsp90 inhibitors, some other drugs may provoke induction of Hsps in target cancer cells. For example, bortezomib (Velcade) is an inhibitor of proteasomes which was found to be effective against multiple myeloma [92]; this anticancer drug, causing intracellular accumulation of aberrant proteins, stimulates the HSF1 activation-dependent transcriptional heat stress response (see Figure 1). It was demonstrated that the bortezomib-stimulated up-regulation of Hsp70 and Hsp27 in MM.1S myeloma cells can impair the expected antitumor effects, while NZ28 (an inhibitor of *Hsp70*- and *Hsp27*-mRNA translation) prevented the Hsp up-regulation and also strengthened apoptosis in the bortezomib-treated cancer cells [89]. The HSF1 activation and up-regulation of Hsps in cancer cells may happen upon administration of a non-steroidal anti-inflammatory drug, indomethacin [93] and, possibly, in some other cases of chemotherapeutic intervention. Certainly, this problem demands development and administration of appropriate inhibitors of the Hsp induction (see Figure 4).

Hsp70

Even without application of Hsp70-inducing drugs, Hsp70 represents by itself a promising target for anticancer chemotherapy. Many human malignancies have the constitutively enhanced expression of an inducible form of Hsp70 that contributes to their resistance to cytotoxic drugs [12, 56]. It was shown on various cancer cell lines that antisense- or siRNA-induced down-regulation of Hsp70 can confer cancer cell sensitization to doxorubicin, cisplatin, vinblastin, taxol, MG132 (an inhibitor of proteasomes) and resveratrol [12, 56, 57, 94, 95].

C. Garrido's research group had used an original approach with targeting Hsp70 by specific peptide molecules which, being introduced into cancer cells, bind Hsp70 and neutralize its antiapoptotic activity [96]. At first, these researchers have designed a peptide construct copying the minimal AIF region required for Hsp70 binding: this AIF derivative, called ADD70, interacts with the peptide-binding domain of Hsp70 and interferes with the interactions of Hsp70 with its client proteins including AIF. In tumor cell lines and in rodent models of cancer, ADD70 sensitized cancer cells to apoptosis induced by doxorubicin or cisplatin [96]. Later, by means of yeast two-hybrid screening of peptide aptamers, this research group has identified the 13-amino acid peptide, called P17, which specifically binds the ATP-binding domain of inducible Hsp70 and inhibits its chaperone activity; in a rodent model, injections of P17 suppressed growth of B16F10 melanoma and sensitized this tumor to cisplatin and 5-ftoruracil [97].

Non-peptide inhibitors of Hsp70 are also known. For example, 2-phenylethynesulfonamide (PES) has been described as a small molecule selectively interacting with Hsp70 and disrupting the association between Hsp70 and its co-chaperones and substrate proteins [98, 99]. *In vitro*, PES promotes cell death in cultured tumor cells by causing protein aggregation and lysosomal/proteasomal dysfunction; *in vivo*, PES was shown to retard lymphoma growth in a murine model [98, 99]. Besides, a novel quinone-based pentacyclic derivative, DTNQ-Pro, has been characterized as an inhibitor of Hsp70 expression and an inducer of apoptosis in human colorectal carcinoma-derived cells that allowed to suggest the potential use of this agent in combined chemotherapy for colon cancer [100].

It should be noted that, as well as in situation with Hsp90 inhibitors (see above), treatments of tumors with inhibitors of the Hsp70 chaperone function may cause the HSF1 activation followed by Hsp induction in the target cancer cells. If so, the problem of inhibitors of HSF1 activation and Hsp expression seems acute again.

Hsp27

As for Hsp27, its down-regulation by means of antisense or siRNA vectors was shown to confer chemosensitization of human lymphoma [101], and breast [102], colon [59], prostate [40], pancreatic [41] and bladder cancer cells [103] that was manifested in the enhanced apoptotic death or senescence and decreased colony-forming ability. Likewise, siRNA-induced depletion of Hsp27 in human erythroleukemic cells increased their sensitivity to etoposide [61]. All these data obtained with using a siRNA-approach characterize Hsp27 as a promising molecular target for enhancing the efficacy of anticancer chemotherapy.

Yet more hopeful results have been obtained with OGX-427, a modified antisense oligonucleotide targeting Hsp27 [41, 103]. So, the OGX-427-induced down-regulation of Hsp27 enhanced sensitivity of human bladder cancer cells to paclitaxel and significantly inhibited the tumor xenograft growth in mice [103]. In another model with pancreas cancer MiaPaCa-2 cells grown in vitro or xenografted in mice, OGX-427 inhibited proliferation and induced apoptosis in the tumor cells and also enhanced their sensitivity to gemcitabine via an elF4E-involving mechanism [41]. At present, Phase II clinical trials examining effects of OGX-427 in various cases of breast, ovarian, bladder, prostate and lung cancer are performed in the US and Canada.

Interestingly, beneficial down-regulation of Hsp27 in cancer cells may be achieved by treatment with KNK437, a known inhibitor of the HSF1-mediated Hsp induction [104]. In a model with gemcitabine-resistant pancreatic tumor cell line, KLM1-R, treatments with KNK437 dramatically reduced the Hsp27 expression level and enhanced the cancer cell sensitivity to gemcitabine [105].

Besides attempts to down-regulate the Hsp27 expression, approaches to inhibition of its functional activity are also considered. In particular, RP101 (Bromovinyldeoxyuridine or "Brivudine") was described as a compound binding Hsp27 and inhibiting interactions of this chaperone with its protein partners [106]. Enhancement of caspase-9 activation followed by enhancement of apoptosis took place in RP101-treated cancer cell cultures. *In vivo*, RP101 improved antitumor effects of chemotherapy in a rat model; in clinical trials with late-stage pancreatic patients (Phase II, NCT00550004), the RP101 dose of 500 mg/day was safe and efficient [106]. Thus, RP101 seems to be an appropriate agent for tumor chemosensitization via direct targeting Hsp27.

An approach with a peptide-based inhibitor, resembling Garrido's approach toward Hsp70 [96], was made by Lee et al. for targeting Hsp27 [107]. Taking into consideration that the protein kinase PKCδ V5 catalytic heptapeptide (FEQFLDI) can directly bind Hsp27 and block its antiapoptotic activity, the Korean researchers had successfully used this PKCδ V5-derived heptapeptide (HEPT) for chemosensitization of human cancer cells. HEPT was shown to sensitize cultured *in vitro* or xenografted in mice human lung carcinoma NCI-H1299 cells to cisplatin; the HEPT-conferred chemosensitization was manifested in enhanced apoptosis and the decreased clonogenicity and delayed *in vivo* tumor growth [107].

In contrast to Hsp90 and Hsp70, Hsp27 is not involved in regulation of HSF1. However, the inhibitor-induced dysfunction of the "small" chaperone may also lead to intracellular accumulation of aberrant proteins that, in turn, may cause the undesirable HSF1-mediated Hsp induction in the treated cancer cells. Obviously, this problem needs further studying.

Targeting Hsps Can Sensitize Tumors to Radiotherapy

Ionizing radiation has generally been accepted as a therapeutic modality in fight against cancer. However, radiation exposure may stimulate radioadaptive selection among the cancer cell variations toward propagation of their radioresistant phenotypes. Malignant cells possessing such phenotypes can endure high-dose radiation and remain proliferating after radiotherapeutic treatments. Moreover, patient's normal tissues also suffer from irradiation. That is why many researchers' efforts are aimed at development of methods of tumor radiosensitization that would enable to decrease the radiation dose-response threshold for malignant tumors without increasing the radiosensitivity of normal cells. In this respect, Hsps seem to be promising molecular targets for elevating the efficacy of anticancer radiotherapy.

Hsp90

As several Hsp90 client proteins contribute to the radioprotective response of cancer cells, inhibitors of Hsp90 may act as radiosensitizers of tumors [11, 15, 49, 50]. Taking into consideration that 17AAG binds to Hsp90 in cancer cells with about 100-fold higher affinity than in normal cells [86], it is expected that this inhibitor will sensitize the target tumors to radiotherapy without aggravating radiation damage to normal tissues.

Indeed, many publications report that Hsp90 inhibitors sensitize tumor cell lines and *in vivo* growing tumor xenografts to radiation (summarized in refs [49, 50]); this was manifested in enhanced cell death, cell cycle arrest and repressed tumor growth. In some of those studies, the radiosensitization taking place in Hsp90 inhibitor-treated cancer cells was not observed in normal cells treated by the same way [108-110]. Although the most of publications describe the radiosensitizing action of geldanamycin and its derivatives 17AAG and 17DMAG, very similar results were obtained following Hsp90 inhibition with radicicol [111], deguelin [112, 113], NVP-AUY922 and NVP-BEP800 [114], BIIB021 (or CNF2024) [115] and a HDAC inhibitor, LBH589 [116]. While hypoxic conditions within solid tumors are known to decrease the efficacy of radiotherapy, the Hsp90 inhibition with NVP-AUY922 was shown to enhance the radiosensitivity of hypoxic tumor cells *in vitro* [117].

At the molecular level, the Hsp90 inhibition-induced radiosensitization of tumor cells was associated with down-regulation of EGF receptor, ErbB2, Raf-1, Akt, Cdc25C and FAK by stimulating their ubiquitin-dependent degradation at proteasomes. Apparently, disruption of ErbB1- , PI3K/Akt- and Raf-1-mediated signaling pathways facilitates the fulfillment of irradiation-triggered apoptotic scenario [11, 15, 50]. A role of Hsp90-HIF1alpha interaction has been revealed in the lung cancer cell radioresistance that was sensitive to 17AAG or deguelin [112].

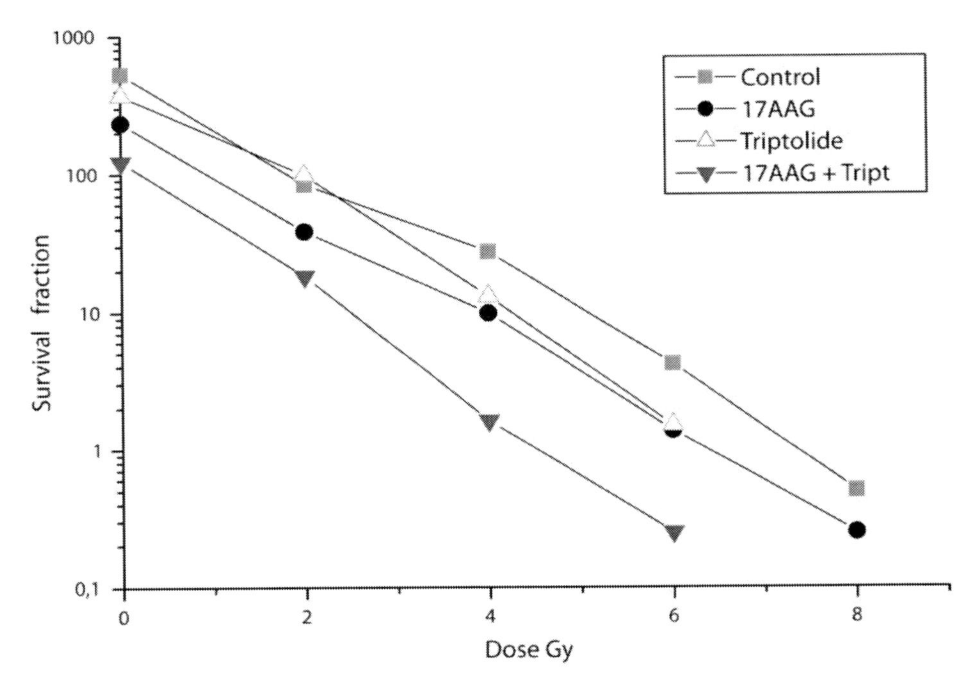

Figure 6. Radiosensitization of human breast cancer MCF-7 cells by inhibiting Hsp90 chaperone activity and/or Hsp induction. The MCF-7 cell cultures were incubated with 35 nM 17AAG (an inhibitor of the Hsp90 activity) and/or 2 nM Triptolide (an inhibitor of the Hsp induction) or without these inhibitors (drug-free control) for 2 hours; then the cells were subjected to low-dose gamma-radiation (2 Gy). Here it is clearly seen that combination of both the inhibitors exerts the much stronger radiosensitizing effect than either inhibitor being used one without the other. This confirms our suggestion that prevention of the Hsp90 inhibition-induced up-regulation of Hsps in tumor cells has to enhance the tumor-sensitizing effects. (V.A. Kudryavtsev, before unpublished data.)

In addition, Hsp90 inhibitors can repress DNA repair in irradiated cancer cells. It was established that the sensitizing pretreatment with 17AAG inhibits the homologous recombination repair of DNA ds breaks in irradiated cells of human prostate carcinoma and lung squamous carcinoma; however, neither the radiosensitization nor the inhibition of DNA repair was observed in normal fibroblasts treated by the same way [110]. Under treatment of human pancreatic carcinoma cell lines, 17DMAG conferred the radiosensitization and inhibited the ds DNA break repair via an impairment of irradiation-responsive DNA-PK catalytic subunit (DNA-PKcs) phosphorylation and a disruption of DNA-PKcs/ErbB1 interaction [47]. In the same work, 17DMAG was shown to abrogate the activation of G_2- and S-phase cell cycle checkpoints that was associated with suppression of irradiation-induced ataxia-telangiectasia mutated (ATM) activation and ATM foci formation in the cell nuclei [47]. In the other study on non-small cell lung cancer cell lines [48], 17DMAG enhanced radiation-induced cytotoxicity by reducing activities of apurinic/apyrimidinic endonuclease and DNA polymerase-beta (both are key enzymes in the base excision repair machinery).

Our own observations have revealed that the Hsp90 inhibition in 17AAG-pretreated, gamma-ray-irradiated human breast cancer cells markedly retards formation and subsequent disappearance of phosphorylated histone gamma H2AX foci (*in situ* sites of DNA ds breakage/repair) in the cell nuclei (Figure 3). This finding suggests that the Hsp90 inhibition with 17AAG interferes with the DNA ds break repair in tumor cells subjected to irradiation.

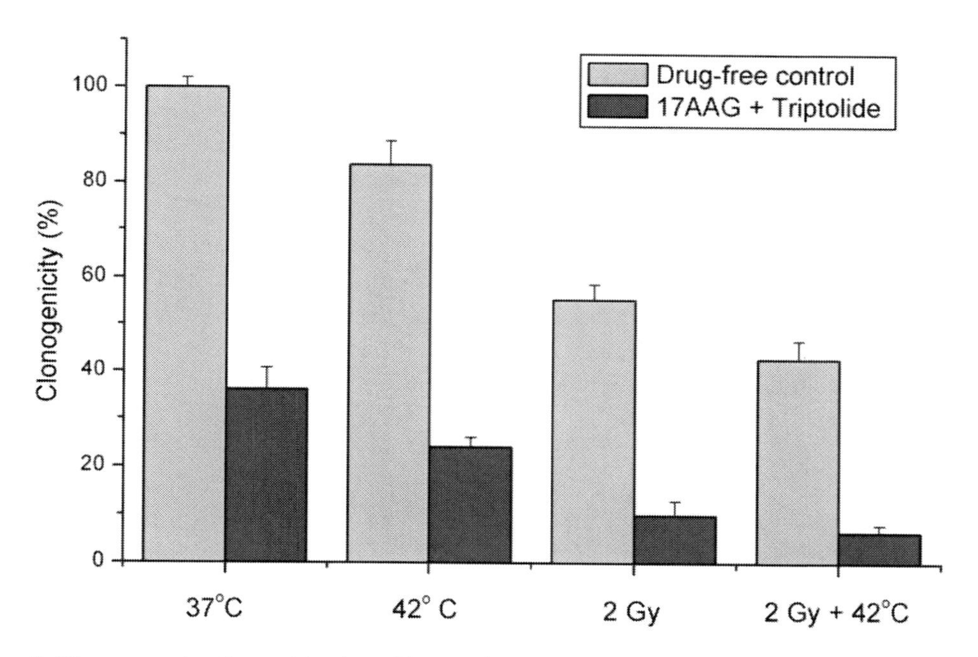

Figure 7. Thermo- and radiosensitization of human breast cancer MCF-7 cells by inhibitors of Hsp90 activity (17AAG) and Hsp induction (Triptolide). The MCF-7 cell cultures were incubated with 35 nM 17AAG and 2 nM Triptolide or without these inhibitors (drug-free control) for 2 hours; then the cells were subjected to mild hyperthermia (42 C, 1 h) and/or low-dose gamma-radiation (2 Gy). Relative levels of the clonogenicity were calculated as percentage from the clonogenicity of the untreated cells. Here it is clearly seen that treatment with both the inhibitors (17AAG + Triptolide) yields 3-6-fold decrease in the cancer cell clonogenicity (see also Figure 6). Such an approach may promote more effective killing of tumor cells by radiotherapy and hyperthermia. (V.A. Kudryavtsev, before unpublished data).

However, because Hsp90 inhibition in cells causes HSF1-mediated induction of cytoprotective Hsp70, Hsp27 and others (see Figure 4 and refs [87-89, 91]), up-regulation of inducible Hsps in target tumors has to impair the radiosenstizing effect of Hsp90 inhibitors. To alleviate this problem we propose to use the Hsp90 inhibitors along with HSF1 inhibitors. We have already proved the potential correctness of such an approach: in our model experiments, triptolide-induced abrogation of the Hsp induction in 17AAG-treated human breast cancer MCF-7 cells did significantly enhance their radiosensitization resulting from the Hsp90 inhibition (Figures 6 and 7).

Hsp70

Being a potent cytoprotector from genotoxic stresses, Hsp70 can contribute to the radioresistance of cancer cells (reviewed in [11, 12]); consequently, inhibition of the expression or functional activity of Hsp70 would sensitize human malignancies to radiotherapy. This is well confirmed by *in vitro* studies on cancer cell cultures. So, using a siRNA approach, R. Bases [54, 55] has shown that down-regulation of Hsp70 in human leukemic cells sensitizes them to radiation via suppression of a suggested Hsp70-involving mechanism of post-radiation DNA repair.

The other researchers described that siRNA-conferred down-regulation of the Hsp70 levels in human prostate carcinoma PC-3 cells or colon carcinoma HCT116 cells causes a decrease in the post-radiation cell survival and enhanced senescence [12, 57, 94]. At the molecular level, this radiosensitization appears to be due to the Hdm2/p53/p21-mediated pathway triggered by the radiation-induced DNA damage response in the Hsp70-depleted cancer cells [12, 57].

So far there are no reports about attempts to radiosensitize tumor cells using small-molecule inhibitors of the Hsp70 activity or expression (e.g. ADD70 and P17 peptides, PES or others), as it was made in the models with chemosensitization (see above). Nevertheless, we suppose that small-molecule Hsp70 inhibitors can also be effective radiosensitizers for *in vivo* targeting malignant tumors and improving the outcome of radiotherapy.

Hsp27

Hsp27 has the powerful cytoprotective potential and can preserve cancer cells from death or senescence following genotoxic stresses (see above and refs [11, 12, 59]). Therefore, this chaperone seems to be an attractive molecular target for sensitizing tumors to radiotherapy. It was already shown that down-regulation of the Hsp27 expression by an antisense cDNA construct sensitizes human prostate carcinoma DU145 cell cultures to gamma-radiation that was manifested in substantially reduced clonogenicity of the irradiated *Hsp27*-cDNA-transfectants [118].

Yet more advanced *in vitro* investigations were performed by Aloy et al. [63] who, using either antisense or RNAi (interfering RNA) approaches, down-regulated the Hsp27 expression in 3 human radioresistant cancer cell lines: head and neck squamous cell carcinoma CQ20B, PC-3 prostate carcinoma and U87 glioblastoma. The data obtained have demonstrated considerable radiosensitization of the Hsp27-depleted tumor cells that was associated with enhanced apoptosis and clonogenic death and also the decreased basal levels of intracellular glutathione [63].

In turn, Hadchity et al. [*35*] successfully accomplished Hsp27 gene silencing by OGX-427, a second-generation antisense oligonucleotide, to overcome the high radioresistance of human head and neck squamous cell carcinoma CQ20B cells. In an *in vitro* part of that study, the down-regulation of Hsp27 in the cultured carcinoma cells significantly enhanced their apoptotic and clonogenic death resulting from radiation exposure, while the cytoprotective Akt activation became impaired. In the *in vivo* experiments with CQ20B carcinoma xenografted in nude mice, combining OGX-427 with local tumor irradiation (5 × 2 Gy) led to a significant regression of the tumors owing to massive apoptosis and decreased levels of glutathione-mediated antioxidant defense in the target cancer cells. Such combination also yielded a suppression of tumor-associated angiogenesis that was resulting from the achieved impairment of Akt activation. Finally, the combined treatment (OGX-427 + irradiation) improved the survival of CQ20B carcinoma-bearing mice and increased their life span without any signs of acute or prolonged toxicity [35].

Certain small molecule inhibitors of the functional activity of Hsp27 are also tested as potential radiosensitizers of tumors. Exploiting the above described heptapeptide (HEPT) which binds and inhibits Hsp27, the same researchers have demonstrated the HEPT-induced radiosensitization of human lung carcinoma NCI-H1299 cells; an increase in apoptosis and a decrease in both *in vitro* clonogenicity and *in vivo* tumor xenograft growth were observed following the combined treatments with HEPT and irradiation [107]. In their next study, the researchers used zerumbone (ZER), a cytotoxic component isolated from *Zingiber zerumbet* Smith [119]. It was found that ZER modifies disulfide (S-S) cross-linking upon the Hsp27 dimerization so that this ZER-produced modification inhibits the binding affinity between Hsp27 and apoptotic effectors such as cytochrom *c* and PKCδ. Pretreatment with ZER prior to radiation exposure conferred radiosensitization of tumor cells cultured *in vitro* and tumor xenografts growing in nude mice [119].

Taken together all these reviewed data afford "proof-of-principle" toward the suggested strategy on targeting Hsps for tumor sensitization. Thus, appropriate (accepted after all clinical trials) inhibitors of the Hsp expression/activity may be used in combination with anticancer radiotherapy.

Targeting HSF1 and Hsps can Sensitize Tumors to Hyperthermia

Hyperthermia remains to be one of clinically applicable methods for elimination of malignant cells. In some cases, hyperthermia can rather effectively kill tumor cells or sensitize them to radiotherapy and chemotherapy [120-123]. Because of the huge progress achieved in development of technical devices for producing total or local hyperthermia into patient's body, the beneficial application of heating as a co-adjuvant in combined treatment of cancer is gaining new prospects. One of the serious drawbacks herein is the heat-induced HSF1 activation and subsequent up-regulation of Hsps in hyperthermia-treated tumors that increases tumor cell survival. In the case of alternate serial treatments, the hyperthermia-provoked increase in Hsp level in cancer cells may render them more resistant to chemo- and radiotherapy (see Figure 4). Therefore, therapeutic strategies aimed at inhibiting HSF1 or Hsps would be useful to thermosensitize cancer cells and enhance the antitumor action of hyperthermia.

It is HSF1 that plays a key role in the stress-responsive mechanism of Hsp induction in mammalian cells (see Introduction and Figure 1). Apparently, artificially arranged dysfunction of HSF1 in cells of malignancies would sensitize them to hyperthermia via abrogation of the *hsp*-gene transcription (Figure 4). That is why HSF1 is considered as one of the most promising targets for overcoming the Hsp-dependent thermoresistance of tumors.

In their *in vitro* studies, Rossi et al. [124] tried to sensitize highly thermoresistant HeLa cell culture derived from human cervical carcinoma by targeting HSF1 with a siRNA vector. It was shown that silencing *HSF1* gene leads to a dramatically increase in the cancer cell sensitivity to "hyperthermochemotherapy" (co-treatment: mild hyperthermia + cisplatin), yielding massive (>95%) apoptosis among the co-treated HeLa cells [124].

Besides, various attempts are made to target HSF1 with small molecule inhibitors. In particular, quercetin and its derivatives and related flavonoids are known as agents suppressing the HSF1 activation and Hsp expression in stressed cells [90]. In cancer-related model systems, quercetin was successfully used as an enhancer of hyperthermia-induced killing of leukemic progenitors [125], prostate cancer cells [126] and Ewing's tumor cells [127]. The results of those studies provide "proof-of-principle" of pharmacological inhibition of HSF1 for repressing tumors; unfortunately, concentrations of quercetin that inhibit HSF1 are too high (> 10 μM) to really be applicable in clinics.

The other well-known inhibitor of HSF1 is KNK437 (N-formyl-3,4-methylenedioxy-benzylidene-gamma-butyrolactam) [104]. This compound was shown to sensitize human colon carcinoma [104], squamous cell carcinoma [128] and prostate carcinoma cells [129] to hyperthermia. In another model study, KNK437 was used for enhancing thermal radiosensitization of human lung adenocarcinoma A549 cells by mild hyperthermia combined with low-dose irradiation [130].

Triptolide, a diterpenoid triepoxide derived from Chinese herb *Tripterygium wilfordii*, was also characterized as a potent functional inhibitor of HSF1 [131]; in the same study, the triptolide-induced inhibition of HSF1 in heat-stressed HeLa cancer cells did confer their thermosensitization that was manifested in the increased cell death after heating. We ourselves achieved substantial thermosensitization of cultured HeLa cells by means of HSF1-inhibiting treatments with triptolide or quercetin; in contrast to quercetin, triptolide exerts its inhibitory and thermosensitizing action being in extremely low (2-5 nM) concentrations (V. Kudryavtsev, unpublished observations).

As no clinically acceptable inhibitors of HSF1 were created so far, further search for the appropriate compounds is still continued. For example, the newly developed beta-alanine-rich linear polyamide was recently described that binds HSE in an unusual 1:1 mode and *in vitro* inhibits HSE-HSF1 interactions [*132*]; it is required, however, to investigate this inhibitor in more detail.

Besides the inhibitors of *hsp*-gene transcription (i.e. inhibitors of HSF1), appropriate inhibitors of *Hsp* mRNA translation could also be used for thermosensitizing tumor cells. So, the above mentioned synthetic inhibitor of the translation of *Hsp70*- and *Hsp27* mRNAs, NZ28, enhanced apoptosis in heat-stressed human colon cancer HCT-116 cells [89].

Nevertheless, it should be noticed that powerful and non-selective inhibitors of the HSF1-mediated heat stress response may fatally decrease the Hsp induction-dependent adaptive resource of patient's normal tissues which are also affected by the cytotoxic action of anticancer therapeutics. The same concern arises toward non-selective inhibitors of the Hsp chaperone function. Therefore, the more rational approaches should be aimed at preferential targeting Hsps in malignant cells with minimizing the chaperone dysfunction in normal cells. One of such approaches may be based on the fact that the ratio inducible Hsp70/constitutive

Hsp70 is much higher in cancer cells than in normal cells, while just cancer cells are "addicted" to inducible Hsp70 and particularly sensitive to its down-regulation [12, 56-58]. It follows from this fact that agents specifically targeting inducible Hsp70 may be more toxic for cancer cells than for normal cells; in this case, the agent-associated cytotoxicity will mostly be produced toward malignancies. Indeed, depletion of inducible Hsp70 in cancer cells treated with an antisense- or siRNA-vector led to their chemo- and thermosensitization and activation of senescence program [12, 56-58, 94]. At present, there are no clinically available agents that would be able to distinguish inducible Hsp70 and constitutive Hsp70 by down-regulating or inhibiting the former without affecting the latter; meanwhile, there is the great need in such tumor-sensitizing agents.

The other approach to selective targeting Hsps in cancer cells may be connected with inhibition of the Hsp90 chaperone activity. On one hand, it has been established that Hsp90 in cancer cells binds 17AAG (and probably some other Hsp90 inhibitors) with ~100-fold higher affinity as compared with Hsp90 in normal cells [86]. On the other hand, Hsp90 is known to be essential for behavior of signaling pathways that ensure proliferation and survival of cancer cells [13-16]. Consequently, in the case of *in vivo* administration, the Hsp90 inhibitors may be redistributed with preferential accumulation within tumor area; if so, the cytotoxic action of co-agents (e.g. hyperthermia or radiation, or another anticancer drug) will be locally enhanced toward malignant cells only. Using Hsp90 inhibition with radicicol, Akimoto et al. [133] have demonstrated the thermosensitization of esophageal tumor cells: treatment with radicicol resulted in blockade of Hsp90-dependent Raf-1, p42/p44 Erk and Akt signaling pathways and intensification of apoptosis following heat stress.

However, because of Hsp90 inhibition (as well as hyperthermia by itself) stimulates Hsp induction in the target cells thus contributing to their survival (see above and refs [87-89, 91]), the tumor-thermosensitizing effect of radicicol (or the other Hsp90 inhibitor) is to be impaired by the newly induced Hsps. In order to escape such complications we propose to combine inhibitors of the Hsp90 activity with inhibitors of the Hsp induction (see Figures 4 and 7). We had already used such an approach for the *in vitro* thermosensitization of human breast cancer MCF-7 cells by means of co-treatment with 17AAG (an inhibitor of the Hsp90 activity) and triptolide (an inhibitor of the Hsp induction) prior to mild hyperthermia (Figure 7). In addition, analogous approach with the two-inhibitor-co-treatment allowed us to sensitize these breast cancer cells to combination of mild hyperthermia and low-dose gamma-irradiation (Figure 7). Taking into consideration that the desirable sensitizing effects were achieved with clinically relevant concentrations of either inhibitor (see Figure 7), we suggest a real possibility of application of our approach in clinics.

With regard to Hsp27, interesting observations were made by Oba and coauthors [134]: interferon (IFN)-gamma suppresses Hsp27 basal transcription and promoter activity so that the combined treatment with hyperthermia and IFN-gamma confers tumor cell thermosensitization in both *in vitro* and *in vivo* models.

Conclusion

Numerous data reviewed in the present chapter justify that Hsps in cancer cells are very attractive targets for anticancer therapy. Indeed, the therapeutic (anti-tumor) potential of various inhibitors and down-regulators of Hsp90, Hsp70 and Hsp27 has been proved in many cancer-related models and tests. The results of most trials indicate that artificially induced inhibition of the Hsp expression or function in tumors will suppress their growth and sensitize them to chemotherapy, radiotherapy, hyperthermia and, probably, immunotherapy.

Unfortunately, so far there are no clinically applicable agents to be effectively used for targeting Hsps in patients' malignancies. Among all screened compounds, certain inhibitors of the Hsp90 activity seem to be the most advanced candidates; however, the cytoprotective Hsp induction concomitant the Hsp90 inhibition is the serious drawback herein. In this respect, a promising strategy may be to combine inhibitors of the Hsp90 activity and inhibitors of the Hsp induction that we had successfully modeled in our own studies (see Figures 4-7). Besides, it seems to us that the directions with OGX-427 (Hsp27 antisense oligonucleotide) and small molecule inhibitors of Hsp27 are very promising as well.

So, in spite of all the reviewed issues, we optimistically believe that in a not so distant future, clinical medicine will get the Hsp inhibition-based tools for effective treatment of cancer.

References

[1] Morimoto, R.I. (1998) Regulation of the heat shock transcriptional response: cross talk between a family of het shock factors, molecular chaperones, and negative regulators. *Genes Dev.* 12, 3788-3796.

[2] Santoro, M.G. (2000) Heat shock factors and the control of stress response. *Biochem. Pharmacol.* 59, 55-63.

[3] Arrigo, A.P. (2005) Heat shock proteins as molecular chaperones. *Med. Sci. (Paris)* 21, 619-625.

[4] Kampinga, H.H. (2006) Chaperones in preventing protein denaturation in living cells and protecting against cellular stress. *Handb. Exp. Pharmacol.* 172, 1-42.

[5] Nollen, E.A.A. & Morimoto, R.I. (2002) Chaperoning signaling pathways: molecular chaperones as stress-sensing 'heat shock' proteins. *J. Cell Sci.* 115, 2809-2816.

[6] Pratt, W.B. & Toft, D.O. (2003) Regulation of signaling protein function and trafficking by the hsp90/hsp70-based chaperone machinery. *Exp. Biol. Med.* 228, 111-133.

[7] Arrigo, A.P., Virot, S., Chaufour, S., Firdaus, W., Kretz-Remy, C., Giaz-Latoud, C. (2005) Hsp27 consolidates intracellular redox homeostasis by upholding glutathione in its reduced form and by decreasing iron intracellular levels. *Antioxid. Redox. Signal* 7, 414- 422.

[8] Takayama, S., Reed, J.C. & Homma, S. (2003) Heat-shock proteins as regulators of apoptosis. *Oncogene* 22, 9041-9047.

[9] Beere, H.M. (2004) "The stress of dying": the role of heat shock proteins in the regulation of apoptosis. *J. Cell Sci.* 117, 2641-2651.

[10] Didelot, C., Schmitt, E., Brunet, M., Maingret, L., Parcellier, A. & Garrido, C. (2006) Heat shock proteins: endogenous modulators of apoptotic cell death. *Handb. Exp. Pharmacol.* 172, 171-198.

[11] Kabakov, A.E. Heat shock proteins against radiation-induced apoptosis: facts, hypotheses, prospects. In: Corvin A, editor. *New Developments in Cell Apoptosis Research.* New York, NY: Nova Science Publishers; 2007, pp. 121-158.
[12] O'Callaghan-Sunol C. & Gabai V.L. (2007) Involvement of heat shock proteins in protection of tumor cells from genotoxic stresses. In: Calderwood S, Sherman MY, Ciocca DR, (eds) *Heat Shock Proteins in Cancer*, Springer, pp. 169-190.
[13] Whitesell, L. & Lindquist, S.L. (2005) HSP90 and the chaperoning of cancer. *Nat. Rev. Cancer* 5, 761-772.
[14] Powers, M.V. & Workman, P. (2006) Targeting of multiple signalling pathways by heat shock protein 90 molecular chaperone inhibitors. *Endocr. Relat. Cancer* 13, Suppl 1, S125-S135.
[15] Kabakov, A.E. Geldanamycin derivatives as promising anticancer drugs: therapy via Hsp90 inhibition. In: Spencer P and Holt W, editors. *Anticancer Drugs. Design, Delivery and Pharmacology.* New York, NY: Nova Science Publishers; 2009, pp. 87-113.
[16] Kanwar, J.R., Kamalapuran, S.K. & Kanwar, R.K. (2011) Targeting surviving in cancer: the cell-signalling perspective. *Drug Discov. Today* 16, 485-494.
[17] Pandey, P., Saleh, A., Nakazawa, A., Kumar, S., Srinivasula, S.M., Kumar, V., Weichselbaum, R., Nalin, C., Alnemri, E.S., Kufe, D., & Kharbanda, S. (2000) Negative regulation of cytochrom c-mediated oligomerization of Apaf-1 and activation of procaspase-9 by heat shock protein 90. *EMBO J.* 19, 4310-4322.
[18] Lewis, J., Devin, A., Miller, A., Lin, Y., Rodrigues, Y., Neckers, L. & Liu, Z.G. (2000) Degradation of the death domain kinase RIP induced by the Hsp90 specific inhibitor geldanamycin sensitizes cells to TNF-induced apoptosis. *J. Biol. Chem.* 275, 10519-10526.
[19] Beere, H.M., Wolf, B.B., Cain, K., Mosser, D.D., Mahboubi, A., Kuwana, T., Tailor, P., Morimoto, R.I., Cohen, G.M. & Green, D.R. (2000) Heat-shock protein 70 inhibits apoptosis by preventing recruitment of procaspase-9 to the Apaf-1 apoptosome. *Nat. Cell Biol.* 2, 469-475.
[20] Saleh, A., Srinivasula, S.M., Balkir, L., Robbins, P.D. & Alnemri, E.S. (2000) Negative regulation of the Apaf-1 apoptosome by Hsp70. *Nat. Cell Biol.* 2, 476-483.
[21] Guo, F., Sigua, C., Bali, P., George, P., Fiskus, W., Skuto, A., Annavarapu, S., Mouttaki, A., Sondarva, G., Wei, S., Wu, J., Djeu, J. & Bhalla, K. (2005) Mechanistic role of heat shock protein 70 in Bcr-Abl-mediated resistance to apoptosis in human acute leukemia cell. *Blood* 105, 1246-1255.
[22] Stankiewicz, A.R., Lachapelle, G., Foo, C.P.Z., Radicioni, S.M. & Mosser, D.D. (2005) Hsp70 inhibits heat-induced apoptosis upstream of mitochondria by preventing Bax translocation. *J. Biol. Chem.* 280, 38729-38739.
[23] Ravagnan, L., Gurbuxani, S., Susin, S.A., Maisse, C., Daugas, E., Zamzami, M., Mak, T., Jaattella, M., Penninger, J.M., Garrido, C. & Kroemer, G. (2001) Heat-shock protein 70 antagonizes apoptosis-inducing factor. *Nat. Cell Biol.* 3, 839-843.
[24] Gabai, V.L., Yaglom, J.A., Volloch, V., Meriin, A.B., Force, T., Koutroumanis, M., Massie, B., Mosser, D.D. & Sherman, M.Y. (2000) Hsp72-mediated suppression of c-Jun N-terminal kinase is implicated in development of tolerance to caspase-independent cell death. *Mol. Cell. Biol.* 20, 6826-6836.

[25] Gabai, V.L., Mabuchi, K., Mosser, D.D. & Sherman, M.Y. (2002) Hsp72 and stress kinase c-jun N-terminal kinase regulate the Bid-dependent pathway in tumor necrosis factor-induced apoptosis. *Mol. Cell. Biol.* 22, 3415-3424.

[26] Meriin, A.B., Yaglom, J.A., Gabai, V.L., Zon, L., Ganiatsas, S., Mosser, D.D., Zon, L. & Sherman, M.Y. (1999) Protein-damaging stresses activate c-Jun N-terminal kinase via inhibition of its dephosphorylation: a novel pathway controlld by HSP72. *Mol. Cell. Biol.* 19, 2547-2555.

[27] Bruey, J.M., Ducasse, C., Bonniaud, P., Ravagnan, L., Susin, S.A., Diaz-Latoud, C., Gurbuxani, S., Arrigo, A.P., Kroemer, G., Solary, E. & Garrido, C. (2000) Hsp27 negatively regulates cell death by interacting with cytochrome c. *Nat. Cell Biol.* 2, 645-652.

[28] Paul, C., Manero, F., Gonin, S., Kretz-Remy, C., Virot, S. & Arrigo, A.-P. (2002) Hsp27 as a negative regulator of cytochrome *c* release. *Mol. Cell. Biol.* 22, 816-834.

[29] Charette, S.J., Lavoie, J.N., Lambert, H. & Lanfry, J. (2000) Inhibition of Daxx-mediated apoptosis by heat shock protein 27. *Mol. Cell. Biol.* 20, 7602-7612.

[30] Hayashi, N., Peacock, J.W., Beraldi, E., Zoubeidi, A., Gleave, M.E. & Ong, C.J. (2011) Hsp27 silencing coordinately inhibits proliferation and promotes Fas induced apoptosis by regulating the PEA-15 molecular switch. *Cell Death Differ.* (in press).

[31] Parcellier, A., Schmitt, E., Gurbuxani, S., Seigneurin-Berny, D., Pance, A., Chantome, A., Plenchette, S., Khochbin, S., Solary, E. & Garrido, C. (2003) HSP27 is a ubiquitin-binding protein involved in I-κBα proteasomal degradation. *Mol. Cell. Biol.* 23, 5790-5802.

[32] Zoubeidi, A., Zardan, A., Wiedmann, R.M., Locke, J., Beraldi, E., Fazli, L. & Gleave, M.E. (2010) Hsp27 promotes insulin-like growth factor-1 survival signaling in prostate cancer via p90Rsk-dependent phosphorylation and inactivation of BAD. *Cancer Res.* 70, 2307-2317.

[33] Rane, M.J., Pan, Y., Singh, S., Powell, D.W., Wu, R., Cummins, T., Chen, Q., McLeish, K.R. & Klein, J.B. (2003) Heat shock protein 27 controls apoptosis by regulating Akt activation. *J. Biol. Chem.* 278, 27828-27835.

[34] Zhang, Y. & Shen, X. (2007) Heat shock protein 27 protects L929 cells from cisplatin-induced apoptosis by enhancing Akt activation and abating suppression of thioredoxin reductase activity. *Clin. Cancer Res.* 13, 2855-2864.

[35] Hadchity, E., Aloy, M.T., Paulin, C., Armandy, E., Watkin, E., Rousson, R., Gleave, M., Chapet, O. & Rodriguez-Lafrasse, C. (2009) Heat shock protein 27 as a new therapeutic target for radiation sensitization of head and neck squamous cell carcinoma. *Mol. Ther.* 17, 1387-1394.

[36] Lee, Y.-J., Lee, D.-H., Cho, C.-K., Bae, S., Jhon, G.-J., Lee, S.-J., Soh, J.-W. & Lee, Y.-S. (2005) HSP25 inhibits protein kinase Cδ–mediated cell death through direct interaction. *J. Biol. Chem.* 280, 18108-18119.

[37] Lee, Y.-J., Lee, D.-H., Cho, C.-K., Chung, H.-Y., Bae, S., Jhon, G.-J., Soh, J.-W., Jeoung, D.-I., Lee, S.-J. & Lee, Y.-S. (2005) HSP25 inhibits radiation-induced apoptosis through reduction of PKCδ–mediated ROS production. *Oncogene* 24, 3715-3725.

[38] Cho, H.N., Lee, Y.J., Cho, C.K., Lee, S.J. & Lee, Y.-S. (2002) Downregulation of ERK2 is essential for the inhibition of radiation-induced cell death in HSP25 overexpressed L929 cells. *Cell Death Differ.* 9, 448-456.

[39] Lee, Y.J., Cho, H.N., Jeoung, D.I., Soh, J.W., Cho, C.K., Bae, S., Chung, H.Y., Lee, S.J. & Lee, Y.S. (2004) HSP25 overexpression attenuates oxidative stress-induced apoptosis: roles of ERK1/2 signaling and manganese superoxide dismutase. *Free Radic. Biol. Med.* 36, 429-444.

[40] Andrieu, C., Taieb, D., Baylot, V., Ettinger, S., Soubeyran, P., De-Thonel, A., Nelson, C., Garrido, C., So, A., Fali, L., Bladou, F., Gleave, M., Iovanna, J.L. & Rocchi, P. (2010) Heat shock protein 27 confers resistance to androgen ablation and chemotherapy in prostate cancer through elF4E. *Oncogene* 29, 1883-1896.

[41] Baylot, V., Andrieu, C., Katsogianou, M., Taieb, D., Garsia, S., Giusiano, S., Acunzo, J., Iovanna, J., Gleave, M. & Rocchi, P. (2011) OGX-427 inhibits tumor progression and enhances gemcitabine chemotherapy in pancreatic cancer. *Cell Death Dis.* 2, e221.

[42] Khanna, K.K. & Jackson, S.P. (2001) DNA double-strand breaks: signaling, repair, and the cancer connection. *Nat. Genet.* 27, 247-254.

[43] Sancar, A., Lindsey-Boltz, L.A., Unsal-Kacmaz, K. & Linn, S. (2004) Molecular mechanisms of mammalian DNA repair and the DNA damage checkpoints. *Annu. Rev. Biochem.* 73, 39-85.

[44] Niida, H. & Nakanishi, M. (2006) DNA damage checkpoints in mammals. *Mutagenesis* 21, 3-9.

[45] Schmitt, C.A. (2007) Cellular senescence and cancer treatment. *Biochim. Biophys. Acta* 1775, 5-20.

[46] Sherman, M., Gabai, V., O'Callaghan, C. & Yaglom, J. (2007) Molecular chaperones regulate p53 and suppress senescence program. *FEBS Lett.* 581, 3711-3715.

[47] Dote, H., Burgan, W.E., Camphausen, K. & Tofilon, P.J. (2006) Inhibition of Hsp90 compromises the DNA damage response to radiation. *Cancer Res.* 66, 9211-9220.

[48] Koll, T.T., Feis, S.S., Wright, M.H., Teniola, M.M., Richardson, M.M., Robles, A.I., Capala, J. & Varticovski, L. (2008) HSP90 inhibitor, DMAG, synergizes with radiation of lung cancer cells by interfering with base excision and ATM-mediated DNA repair. *Mol. Cancer Ther.* 7, 1985-1992.

[49] Camphausen, K. & Tofilon, P.J. (2007) Inhibition of Hsp90: a multitarget approach to radiosensitization. *Clin. Cancer Res.* 13, 4326-4330.

[50] Kabakov, A.E., Kudryavtsev, V.A. & Gabai, V.L. (2010) Hsp90 inhibitors as promising agents for radiotherapy. *J. Mol. Med.* 88, 241-247.

[51] Sharma, K., Vabulas, R.M., Macek, B., Pinkert, S., Cox, J., Mann, M. & Hartl, F.U. (2012) Quantitative proteomics reveals that Hsp90 inhibition preferentially targets kinases and the DNA damage response. *Mol. Cell. Proteomics* 11, M111.014654.

[52] Kenny, M.K., Mendez, F., Sandigursky, M., Kureekattil, R.P., Goldman, J.D., Franklin, W.A. & Bases, R. (2001) Heat shock protein 70 binds to human apurinic/apyrimidinic endonuclease and stimulates endonuclease activity at abasic sites. *J. Biol. Chem.* 276, 9532-9536.

[53] Niu, P., Gong, Z., Tan, H., Wang, F., Yuan, J., Feng, Y., Wei, Q., Tanguay R.M. & Wu, T. (2006) Overexpressed heat shock protein 70 protects cells against DNA damage caused by ultraviolet C in a dose-dependent manner. *Cell Stress Chaperones* 11, 162-169.

[54] Bases, R. (2005) Clonogenicity of human leukemic cells protected from cell-lethal agents by heat shock protein 70. *Cell Stress Chaperones* 10, 37-45.

[55] Bases, R. (2006) Heat shock protein 70 enhanced deoxyribonucleic acid base excision repair in human leukemic cells after ionizing radiation. *Cell Stress Chaperones* 11, 240-249.

[56] Yaglom, J.A., Gabai, V.L., Sherman, M.Y. (2007) High levels of heat shock protein hsp72 in cancer cells suppress default senescence pathways. *Cancer Res.* 67, 2373-2381.

[57] Gabai, V.L., Sherman, M.Y. & Yaglom, J.A. (2010) Hsp72 depletion suppresses gammaH2AX activation by genotoxic stresses via p53/p21 signaling. *Oncogene* 29, 1952-1962.

[58] Gabai, V.L., Yaglom, J.A., Waldman, T. & Sherman, M.Y. (2009) Heat shock protein Hsp72 controls oncogene-induced senescence pathways in cancer cells. *Mol. Cell. Biol.* 29, 559-569.

[59] O'Callaghan-Sunol, C., Gabai, V.L., Sherman, M.Y. (2007) Hsp27 modulates p53 signaling and suppresses cellular senescence. *Cancer Res.* 67, 11779-11788.

[60] Kanagasabai, R., Krishnamuthy, K., Vedam, K., Wang, Q., Zhu, Q. & Ilangovan, G. (2010) Hsp27 protects adenocarcinoma cells from UV-induced apoptosis by Akt and p21 dependent pathway of survival. *Mol. Cancer Res.* 8, 1399-1412.

[61] Schepers, H., Geugien, M., van der Toorn, M., Bryantsev, A.L., Kampinga, H.H. & Vellenga, E. (2005) HSP27 protects AML cells against VP-16-induced apoptosis through modulation of p38 and c-Jun. *Exp. Hematol.* 33, 660-670.

[62] Park, Y.M., Han, M.Y., Blackburn, R.V. & Lee, Y.J. (1998) Overexpression of HSP25 reduces the level of TNF alpha-induced oxidative DNA damage biomarker, 8-hydroxyguanosine, in L929 cells. *J. Cell. Physiol.* 174, 27-34.

[63] Aloy, M.T., Hadchitity, E., Bionda, C., Diaz-Latoud, C., Claude, L., Rousson, R., Arrigo, A.P. & Rodriguez-Lafrasse, C. (2008) Protective role of Hsp27 against gamma radiation-induced apoptosis and radiosensitization effects of Hsp27 gene silencing in different human tumor cells. *Int. J. Radiat. Oncol. Biol. Phys.* 70, 543-553.

[64] Yi, M.J., Park, S.H., Cho, H.N., Yong Chung, H., Kim, J.I., Cho, C.K., Lee, S.J. & Lee, Y.S. (2002) Heat-shock protein 25 (Hspb1) regulates manganese superoxide dismutase through activation of Nfkb (NF-kappaB). *Radiat. Res.* 158, 641-649.

[65] Hwang, M., Moretti, L. & Lu, B. (2009) Hsp90 inhibitors: multi-target antitumor effects and novel combinatorial therapeutic approaches in cancer therapy. *Curr. Med. Chem.* 16, 3081-3092.

[66] Jhaveri, K., Taldone, T., Modi, S. & Chiosis, G. (2012) Advances in the clinical development of heat shock protein 90 (Hsp90) inhibitors in cancer. *Biochim. Biophys. Acta* 1823, 742-755.

[67] Whitesell, L. & Linn, N.U. (2012) Hsp90 as a platform for the assembly of more effective cancer chemotherapy. *Biochim. Biophys. Acta* 1823, 756-766.

[68] Hawkins, L.M., Jayanthan, A.A. & Narendran, A. (2005) Effects of 17-allylamino-17-demethoxygeldanamycin (17-AAG) on pediatric acute lymphoblastic leukemia (ALL) with respect to Bcr-Abl status and imatinib mesylate sensitivity. *Pediatr. Res.* 57, 430-437.

[69] Solit, D.B., Basso, A.D., Olshen, A.B., Scher, H.I. & Rosen, N. (2003) Inhibition of heat shock protein 90 function down-regulates Akt kinase and sensitizes tumors to Taxol. *Cancer Res.* 63, 2139-2144.

[70] Sawai, A., Chandarlapaty, S., Greulich, H., Gonen, M., Ye, Q., Arteaga, C.L., Sellers, W., Rosen, N. & Solit, D.B. (2008) Inhibition of Hsp90 down-regulates mutant epidermal growth factor receptor (EGFR) expression and sensitizes EGFR mutant tumors to paclitaxel. *Cancer Res.* 68, 589-596.

[71] Ramalingam, S.S., Egorin, M.J., Ramanathan, R.K., Remick, S.C., Sikorski, R.P., Lagattuta, T.F., Chatta, G.S., Friedland, D.M., Stoller, R.G., Potter, D.M., Ivy, S.P. & Belani, C.P. (2008) A phase I study of 17-allylamino-17- demethoxygeldanamycin combined with Paclitaxel in patients with advanced solid malignancies. *Clin. Cancer Res.* 14, 3456-3461.

[72] Vasilevskaya, I.A., Rakitina, T.V. & O'Dwyer, P.J. (2004) Quantitative effects on c-Jun N-terminal protein kinase signaling determine synergistic interaction of cisplatin and 17-allylamino-17-demethoxygeldanamycin in colon cancer cell lines. *Mol. Pharmacol.* 65, 235-243.

[73] Moser, C., Lang, S.A., Kainz, S., Gaumann, A., Fichtner-Feigl, S., Koehl, G.E., Schlitt, H.J., Geissler, E.K. & Stoeltzing, O. (2007) Blocking heat shock protein-90 inhibits the invasive properties and hepatic growth of human colon cancer cells and improves the efficacy of oxaliplatin in p53-deficient colon cancer tumors in vivo. *Mol. Cancer Ther.* 6, 2868-2878.

[74] Banerji, U., Sain, N., Sharp, S.Y., Valenti, M., Asad, Y., Ruddle, R., Raynaud, F., Walton, M., Eccles, S.A., Judson, I., Jackman, A.L. & Workman, P. (2008) An in vitro and in vivo study of the combination of the heat shock protein inhibitor 17-allylamino-17-demethoxygeldanamycin and carboplatin in human ovarian cancer models. *Cancer Chemother. Pharmacol. 62,* 769-778.

[75] Georgakis, G.V., Li, Y., Rassidakis, G.Z., Medeiros, L.J. & Younes, A. (2006) The HSP90 inhibitor 17-AAG synergizes with doxorubicin and U0126 in anaplastic large cell lymphoma irrespective of ALK expression. *Exp. Hematol.* 34, 1670-1679.

[76] Robles, A.I., Wright, M.H., Gandhi, B., Feis, S.S., Hanigan, C.L., Wiestner, A. & Varticovski, L. (2006) Schedule-dependent synergy between the heat shock protein 90 inhibitor 17-(dimethylaminoethylamino)-17-demethoxygeldanamycin and doxorubicin restores apoptosis to p53-mutant lymphoma cell lines. *Clin. Cancer Res.* 12, 6547-6556.

[77] Yao, Q., Weigel, B. & Kersey, J. (2007) Synergism between etoposide and 17-AAG in leukemia cells: critical roles for Hsp90, FLT3, topoisomerase II, Chk1, and Rad51. *Clin. Cancer Res.* 13, 1591-1600.

[78] Premkumar, D.R., Arnold, B. & Pollack, I.F. (2006) Cooperative inhibitory effect of ZD1839 (Iressa) in combination with 17-AAG on glioma cell growth. *Mol. Carcinog.* 45, 288-301.

[79] Radujkovic, A., Schad, M., Topaly, J., Veldwijk, M.R., Laufs, S., Schultheis, B.S., Jauch, A., Melo, J.V., Fruehauf, S. & Zeller, W.J. (2005) Synergistic activity of imatinib and 17-AAG in imatinib-resistant CML cells overexpressing BCR-ABL--Inhibition of P-glycoprotein function by 17-AAG. *Leukemia* 19, 1198-1206.

[80] Hawkins, L.M., Jayanthan, A.A. & Narendran, A. (2005) Effects of 17-allylamino-17-demethoxygeldanamycin (17-AAG) on pediatric acute lymphoblastic leukemia (ALL) with respect to Bcr-Abl status and imatinib mesylate sensitivity. *Pediatr. Res.* 57, 430-437.

[81] Duus, J., Bahar, H.I., Venkataraman, G., Ozpuyan, F., Izban, K.F., Al-Masri, H., Maududi, T., Toor, A. & Alkan, S. (2006) Analysis of expression of heat shock protein-90 (HSP90) and the effects of HSP90 inhibitor (17-AAG) in multiple myeloma. *Leuk. Lymphoma* 47, 1369-1378.

[82] Mitsiades, C.S., Mitsiades, N.S., McMullan, C.J., Poulaki, V., Kung, A.L., Davies, F.E., Morgan, G., Akiyama, M., Shringarpure, R., Munshi, N.C., Richardson, P.G., Hideshima, T., Chauhan, D., Gu, X., Bailey, C., Joseph, M., Libermann, T.A., Rosen, N.S. & Anderson, K.C. (2006) Antimyeloma activity of heat shock protein-90 inhibition. *Blood* 107, 1092-1100.

[83] Raja, S.M., Clubb, R.J., Bhattacharyya, M., Dimri, M., Cheng, H., Pan, W., Ortega-Cava, C., Lakku-Reddi, A., Naramura, M., Band, V. & Band, H. (2008) A combination of trastuzumab and 17-AAG induces enhanced ubiquitinylation and lysosomal pathway-dependent ErbB2 degradation and cytotoxicity in ErbB2- overexpressing breast cancer cells. *Cancer Biol. Ther.* 7, 1630-1640.

[84] Modi, S., Stopeck, A.T., Gordon, M.S., Mendelson, D., Solit, D.B., Bagatell, R., Ma, W., Wheler, J., Rosen, N., Norton, L., Cropp, G.F., Johnson, R.G., Hannah, A.L. & Hudis, C.A. (2007) Combination of trastuzumab and tanespimycin (17-AAG, KOS-953) is safe and active in trastuzumab-refractory HER-2 overexpressing breast cancer: a phase I dose-escalation study. *J. Clin. Oncol.* 25, 5410-5417.

[85] Bertram, J, Palfner, K, Hiddemann, W. & Kneba, M. (1996) Increase of P-glycoprotein-mediated drug resistance by hsp90 beta. *Anticancer Drugs* 7, 838-845.

[86] Kamal, A., Thao, L., Sensintaffar, J., Zhang, L, Boehm, M.F., Fritz, L.C., and Burrows, F.J. (2003) A high-affinity conformation of Hsp90 confers tumour selectivity on Hsp90 inhibitors. *Nature* 425, 407-410.

[87] Bagatell, R., Paine-Murrieta, G.D., Taylor, C.W., Pulcini, E.J., Akinaga, S., Benjamin, I.J. & Whitesell, L. (2000) Induction of a heat shock factor 1- dependent stress response alters the cytotoxic activity of Hsp90-binding agents. *Clin. Cancer Res.* 6, 3312-3318.

[88] Guo, F., Rocha, K., Bali, P., Pranpat, M., Fiskus, W., Boyapalle, S., Kumaraswamy, S., Balasis, M., Greedy, B., Armitage, E.S.M., Lawrence, N. & Bhalla, K. (2005) Abrogation of heat shock protein 70 induction as a strategy to increase antileukemia activity of heat shock protein 90 inhibitor 17-allylamino-demethoxy geldanamycin. *Cancer Res.* 65, 10536-10544.

[89] Zaarur, N., Gabai, V.L., Porco, J.A., Calderwood, S. & Sherman, M.Y. (2006) Targeting heat shock response to sensitize cancer cells to proteasome and Hsp90 inhibitors. *Cancer Res.* 66, 1783-1791.

[90] Hosokawa, N., Hirayoshi, K., Kudo, H., Takeshi, H., Aoike, A., Kawai, A. & Nagata, K. (1992) Inhibition of the activation of heat shock factor *in vivo* and *in vitro* by flavonoids. *Mol. Cell. Biol.* 12, 3490-3498.

[91] McCollum, A.K., Teneyck, C.J., Sauer, B.M., Toft, D.O. & Erlichman, C. (2006) Up-regulation of heat shock protein 27 induces resistance to 17-allylamino-demethoxygeldanamycin through a glutathione-mediated mechanism. *Cancer Res.* 66, 10967-10975.

[92] Chauhan, D., Hideshima, T., Mitciades, C., Richardson, P. & Anderson, K.S. (2005) Proteasome inhibitor therapy in multiple myeloma. *Mol. Cancer Ther.* 4, 686-692.

[93] Lee, B.S., Chen, J., Angelidis, C., Jurivich, D.A. & Morimoto, R.I. (1995) Pharmacological modulation of heat shock factor 1 by anti-inflammatory drugs results

in protection against stress-induced cellular damage. *Proc. Natl. Acad. Sci. USA* 92, 7207-7211.

[94] Gabai, V.L., Budagova, K.R. & Sherman, M.Y. (2005) Increased expression of the major heat shock protein Hsp72 in human prostate carcinoma cells is dispensable for their viability but confers resistance to a variety of anticancer agents. *Oncogene* 24, 3328-3338.

[95] Zhu, Q., Xu, Y.-M., Wang, L.-F., Zhang, Y., Wang, F., Zhao, J., Jia, L.-T., Zhang, W.-G. & YangA.-G. (2009) Heat shock protein 70 silencing enhances apoptosis inducing factor-mediated cell death in hepatocellular carcinoma HepG2 cells. *Cancer Biol. Ther.* 8:9, 792-798.

[96] Schmitt, E., Maingret, L., Puig, P.E., Rerole, A.L., Ghiringhelli, F., Hammann, A., Solary, E., Kroemer, G. & Garrido, C. (2006) Heat shock protein 70 neutralization exerts potent antitumor effects in animal models of colon cancer and melanoma. *Cancer Res.* 66, 4191-4197.

[97] Rerole A.L., Gobbo, J., De Thonel, A., Schmitt, E., de Barros, J.P.P., Hammann, A., Lanneau, D., Fourmaux, E., Deminov, O., Micheau, O., Lagrost, L., Colas, P., Kroemer, G. & Garrido, C. (2011) Peptides and aptamers targeting HSP70: a novel approach for anticancer therapy. *Cancer. Res.* 71, 484-495.

[98] Leu, J.I., Pimkina, J., Frank, A., Murphy, M.E. & George, D.L. (2009) A small molecule inhibitor of inducible heat shock protein 70 (HSP70). *Mol. Cell* 36, 15-27.

[99] Leu, J.I., Pimkina, J., Pandey, P., Murphy, M.E. & George, D.L. (2011) HSP70 inhibition by the small-molecule 2-phenylethynesulfonamide impairs protein clearance pathways in tumor cells. *Mol. Cancer Res.* 9, 936-947.

[100] Gomez-Monterrey, I., Campiglia, P., Bertamino, A., Aquino, C., Sala, M., Grieco, P., Dicitore, A., Vanacore, D., Porta, A., Maresca, B., Novellino, E. & Stiuso, P. (2010) A novel quinine-based derivative (DTNQ-Pro) induces apoptotic death via modulation of heat shock protein expression in Caco-2 cells. *Br. J. Pharmacol.* 160, 931-940.

[101] Chauhan, D., Li, G., Shringarpure, R., Podar, K., Ohtake, Y., Hideshima, T. & Anderson, K.C. (2003) Blockade of Hsp27 overcomes Bortezomib/proteasome inhibitor PS-341 resistance in lymphoma cells. *Cancer Res.* 63, 6174-6177.

[102] Oesterreich, S., Weng, C.N., Qiu, M., Hilsenbeck, S.G., Osborne, C.K. & Fuqua S.A. (1993) The small hat shock protein hsp27 is correlated with growth and drug resistance in human breast cancer cells. *Cancer Res.* 53, 4443-4448.

[103] Kamada, M., So, A., Muramaki, M., Rocchi, P., Beralsi, E. & Gleave, M. (2007) Hsp27 knockdown using nucleotide-based therapies inhibit tumor growth and enhance chemo-therapy in human bladder cancer cells. *Mol. Cancer Ther.* 6, 299-308.

[104] Yokota, S., Kitahara, M. & Nagata, K. (2000) Benzylidene lactam compound, KNK437, a novel inhibitor of acquisition of thermotolerance and heat shock protein induction in human colon carcinoma cells. *Cancer Res.* 60, 2942-2948.

[105] Taba, K., Kuramitsu, Y., Ryozama, S., Yoshida, K., Tanaka, T., Mori-Iwamoto, S., Maehara, S., Maehara, Y., Sakaida, I & Nakamura, K. (2011) KNK437 downregulates heat shock protein 27 of pancreatic cancer cells and enhances the cytotoxic effect of gemcitabine. *Chemotherapy* 57, 12-26.

[106] Heinrich, J.C., Tuukkanen, A., Schroeder, M., Fahrig, T. & Fahrig, R. (2011) RP101 (brivudine) binds to heat shock protein HSP27 (HSPB1) and enhances survival in animals and pancreatic cancer patients. *J. Cancer Res. Clin. Oncol.* 137, 1349-1361.

[107] Lee, H.J., Kim, E.H., Seo, W.D., Choi, T.H., Cheon, G.J., Lee, Y.J. & Lee, Y.S. (2011) Heat shock protein 27-targeted hetapeptide of the PKCdelta catalytic V5 region sensitizes tumors with radio- and chemoresistance. *Int. J. Radiat. Oncol. Biol. Phys.* 80, 221-230.

[108] Russel, J.S., Burgan, W., Oswald, K.A., Camphausen, K. & Tofilon, P.J. (2003) Enhanced cell killing induced by the combination of radiation and the heat shock protein 90 inhibitor 17-allylamino-17-demethoxygeldanamycin: a multitarget approach to radiosensitization. *Clin. Cancer Res.* 9, 3749-3755.

[109] Bisht, K.S., Bradbury, C.M., Mattson, D., Kaushal, A., Sowers, A., Markovina, S., Ortiz, K.L., Sieck, L.K., Isaacs, J.S., Brechbiel, M.W., Mitchell, J.B., Neckers, L.M. & Gius, D. (2003) Geldanamycin and 17-allylamino-17-demethoxy- geldanamycin potentiate the *in vitro* and *in vivo* radiation response of cervical tumor cells via the heat shock protein 90-mediated intracellular signaling and cytotoxicity. *Cancer Res.* 63, 8984-8995.

[110] Noguchi, M., Yu., D., Hirayama, R., Ninomiya, Y., Sekine, E., Kubota, N., Ando, K. & Okayasu, R. (2006) Inhibition of homologous recombination repair in irradiated tumor cells pretreated with Hsp90 inhibitor 17-allylamino-17-demethoxy- geldanamycin. *Biochem. Biophys. Res. Commun.* 351, 658-663.

[111] Harashima, K., Akimoto, T., Nonaka, T., Tsuzuki, K., Mitsuhashi, N. & Nakano, T. (2009) Heat shock protein 90 (Hsp90) chaperone complex inhibitor, radicicol, potentiated radiation-induced cell death in a hormone-sensitive prostate cancer cell line trough degradation of the androgen receptor. *Int. J. Radiat. Biol.* 81, 63-76.

[112] Kim, W.Y., Oh, S.H., Woo, J.K., Hong, W.K. & Lee, H.Y. (2009) Targeting heat shock protein 90 overrides the resistance of lung cancer cells by blocking radiation- induced stabilization of hypoxia-inducible factor-1α. *Cancer Res.* 69, 1624-1632.

[113] Yi, T., Li, H., Wang, X. & Wu, Z. (2008) Enhancement radiosensitization of breast cancer cells by deguelin. *Cancer Biother. Radiopharm.* 23, 355-362.

[114] Stingl, L., Stuhmer, T., Chatterjee, M., Jensen, M.R., Flentje, M. & Djuzenova, C.S. (2010) Novel HSP90 inhibitors, NVP-AUY922 and NVP-BEP800, radiosensitise tumour cells through cell-cycle imparment, increased DNA damage and repair protraction. *Br. J. Cancer* 102, 1578-1591.

[115] Yin, X., Zhang, H., Lundgren, K., Wilson, L., Burrows, F., Shores, C.G. (2009) BIIB021, a novel Hsp90 inhibitor, sensitizes head and neck squamous cell carcinoma to radiotherapy. *Int J Cancer* 126, 1216-1225.

[116] Kim, I.A., No, M., Lee, J.M., Shin, J.H., Oh, J.S., Choi, E.J., Kim, I.H., Atadja, P. & Bernhard, E.J. (2009) Epigenetic modulation of radiation response in human cancer cells with activated EGFR or HER-2 signaling: Potential role of histone deacetylase 6. *Radiother. Oncol.* 92, 125-132.

[117] Djuzenova, C.S., Blassl, C., Rollof, K., Kuger, S., Katzer, A., Niewidok, N., Gunther, N., Polat, B., Sukhorukov, V. & Flentje, M. (2012) Hsp90 inhibitor NVP-AUY922 Enhances radiation sensitivity of tumor cell lines under hypoxia. *Cancer Biol. Ther.* 13, 425-434.

[118] Teimourian, S., Jalal, R., Sohrabpour, M. & Goliaei, B. (2006) Down-regulation of Hsp27 radiosensitizes human prostate cancer cells. *Int. J. Urol.* 13, 1221-1225.

[119] Choi, S.H., Lee, Y.J., Seo, W.D., Lee, H.J., Nam, J.W., Lee, Y.J., Kim, J., Seo, E.K. & Lee, Y.S. (2011) Altered cross-linking of HSP27 by zerumbone as a novel strategy for

overcoming HSP27-mediated radioresistance. *Int. J. Radiat. Oncol. Biol. Phys.* 79, 1196-1205.

[120] Wust, P., Hildebrandt, B., Sreenivasa, G., Rau, B., Gelermann, J., Riess, H., Felix, R. & Schlag P.M. (2002) Hyperthermia in combined treatment of cancer. *Lancet Oncol.* 3, 487-497.

[121] Takahashi, I., Emi, Y., Hasuda, S., Kakeji, Y., Naehara, Y. & Sugimachi, K. (2002) Clinical application of hyperthermia combined with anticancer drugs for the treatment of solid tumors. *Surgery* 131, S78-S84.

[122] Westermann, A.M., Jones, E.L., Schem, B.C., van der Steen-Banasik, E.M., Koper, P., Mella, O., Uitterhoeve, O.L., de Wit, R., van der Velden, J., Burger, C., van der Wilt, (2005) First results of triple-modality treatment combining radiotherapy, chemotherapy, and hyperthermia for the treatment of patients with stage IIB, III, and IVA cervical carcinoma. *Cancer* 104, 763-770.

[123] Kampinga, H.H. (2006) Cell biological effects of hyperthermia alone or combined with radiation or drugs: a short introduction to newcomers in the field. *Int. J. Hyperthermia* 22, 191-196.

[124] Rossi, A., Ciafre, S., Balsamo, M., Pierimatchi, P. & Santoro, G. (2006) Targeting the Heat Shock Factor 1 by RNA interference: a potent tool to enhance hyperthermochemo-therapy efficacy in cervical cancer. *Cancer Res.* 66, 7678-7685.

[125] Larocca, L.M., Ranelletti, F.O., Maggiano, N., Rutella, S., La Barbers, E.O., Rumi, C., Serra, F., Voso, M.T., Piantelli, M., Teofili, L. & Leone, G. (1997) Differential sensitivity of leukemic and normal hematopoietic progenitors to the killing effect of hyperthermia and quercetin used in combination: role of heat-shock protein-70. *Int. J. Cancer* 73, 75-83.

[126] Asea, A., Ara, G., Teicher, B.A., Stevenson, M.A. & Calderwood, S.K. (2001) Effects of the flavonoid drug quercetin on the response of human prostate tumours to hyperthermia in vitro and in vivo. *Int. J. Hyperthermia* 17, 347-356.

[127] Debes, A., Oerding, M., Willers, R., Gobel, U. & Wessalowski, R. Sensitization of Human Ewing's tumor cells to chemotherapy and heat treatment by the bioflavonoid Quercetin. *Anticancer Res.* 23, 3359-3366.

[128] Ohnishi, K., Takahashi, A., Yokota, S. & Ohnishi, T. (2004) Effects of heat shock protein inhibitor KNK437 on heat sensitivity and heat tolerance in human squamous cell carcinoma cell lines differing in p53 status. *Int. J. Radiat. Biol.* 80, 607-614.

[129] Sahin, E., Sahin, M., Sanlioglu, A.D. & Gumuslu, S. (2011) KNK437, a benzylidene lactam compound, sensitises prostate cancer cells to the apoptotic effect of hyperthermia. *Int. J. Hyperthermia* 27, 63-73.

[130] Sakurai, H., Kitamoto, Y., Saitoch, J., Nonaka, T., Ishikawa, H., Kiyohara, H., Shiova, M. Fukushima, M., Akimoto, T., Hasegawa, M. & Nakano, T. (2005) Attenuation of chronic thermotolerance by KNK437, a benzylidene lactam compound, enhances thermal radiosensitization in mild temperature hyperthermia combined with low dose-rate radiation. *Int. J. Radiat. Biol.* 81, 711-718.

[131] Westerheide, S.D., Kawahara, T.L.A., Orton, K. & Morimoto, R.I. (2006) Triptolide, an inhibitor of the human heat shock response that enhances stress-induced cell death. *J. Biol. Chem.* 281, 9616-9622.

[132] Wang, R.E., Pandita, R.K., Cai, J., Hunt, C.R. & Taylor, J.S. (2012) Inhibition of heat shock transcription factor binding by a linear polyamide binding in an unusual 1:1 mode. *Chembiochem.* 13, 97-104.

[133] Akimoto, T., Nonaka, T., Harashima, K., Sakurai, H., Ishikawa, H. & Mitsuhashi, N. (2004) Radicicol potentiates heat-induced cell killing in a human oesophageal cancer cell line: the Hsp90 chaperone complex as a new molecular target for enhancement of thermosensitivity. *Int. J. Radiat. Biol.* 80, 483-492.

[134] Oba, M., Yano, S., Shuto, T., Suico, M.A., Eguma, A. & Kai, H. (2008) IFN-gamma down-regulates Hsp27 and enhances hyperthermia-induced tumor cell death in vitro and tumor suppression in vivo. *Int. J. Oncol.* 32, 1317-1324.

In: Heat Shock Proteins
Editor: Saad Usmani

ISBN: 978-1-62417-571-8

Chapter III

Involvement of Heat Shock Proteins in Normal Follicular Growth and Ovarian Follicular Cysts

Melisa M. L. Velázquez[1,2], Natalia R. Salvetti[1,2], Florencia Rey[1,2], Fernanda M. Rodríguez[1,2], Valentina Matiller[1], M. Eugenia Baravalle[1] and Hugo H. Ortega*[1,2]

[1] Department of Morphological Sciences, Faculty of Veterinary Sciences, National University of Litoral, Argentina

[2] National Scientific and Technical Research Council (CONICET), Argentina

Abstract

To be successful, reproductive processes should be in harmony and synchronized with the environment. In fact, adequate conditions promote reproduction via hormonal stimulators. Adverse conditions, such as inadequate temperature, nutritional deficiencies, and infections, induce stress and suppress reproduction through stress-related substances. These anomalies induce changes in the expression of numerous genes, including genes encoding heat shock proteins (HSPs), which could potentially oppose and even spoil hormonal effects and reproductive success. Moreover, the expression of these genes would be part of the functional response to the hormones and neurotransmitters induced by stress. There are reports indicating that HSPs can control hormonal functions and vice versa. It has been found that the expression of these proteins is higher in the female reproductive organs and that their main functions are related to the maintenance of the configuration of non-stimulated steroid receptors and the modulation of binding to receptor. Furthermore, since the ovarian cycle is associated with changes in the expression of HSPs, these proteins could be in close relationship with anovulatory disorders like cystic ovarian disease (COD) in cattle, where a combination of weak proliferation indices and low apoptosis has been described. In this sense, it has been demonstrated that HSPs are also involved in cell survival and death mechanisms. Therefore, previous observations of aberrant HSPs gene expression in cells of cystic

* hhortega@fcv.unl.edu.ar.

follicles suggest that these proteins could be associated with the intra-ovarian component of COD pathogenesis, due to a link with intracellular apoptosis signaling pathways.

1. Introduction

Heat Shock Proteins (HSPs) form a diverse group of proteins that are classified according to their molecular weight. Most members of the HSP family perform a chaperone or chaperon-like function inside the cell, helping proteins in the crowded intracellular environment to reach the appropriate, final folding state and avoiding folding structures that are not functional and could lead to proteosomal degradation or protein denaturation (Ellis, 1993; Welch, 1993). This chaperone function results in a protective effect of HSPs against noxious damage from heat ischemia or exposure to toxic chemicals.

Efforts from a large number of investigators have shown that the heat shock response is ubiquitous and highly conserved—in all organisms from bacteria to plants and animals—and constitutes an essential defense mechanism for protection of cells from a wide range of harmful conditions, including heat shock, alcohols, inhibitors of energy metabolism, heavy metals, oxidative stress, fever, and inflammation (Lindquist, 1986; Morimoto, 1993).

Among the members of the HSP family of higher molecular mass (HSP105/110), the 105-kDa protein (Hsp105) is one of the most important mammalian HSPs. Zhang et al. (2005) demonstrated that Hsp105 is expressed in the monkey testis and may play an important role in the regulation of germ cell apoptosis induced by heat stress. In mammals, Hsp105 may function as a pro-apoptotic factor or as an anti-apoptotic factor depending on the cell type (Hatayama et al., 2001; Yamagishi et al., 2002). Recent studies showing dynamic changes in the level of expression of Hsp105 during early implantation suggest involvement of this protein in certain cellular events in luminal epithelial and stromal cells, which are essential for implantation (Yuan et al., 2009).

Prominent members of the HSP90 family of proteins are the two Hsp90 isoforms (Hsp90a and Hsp90b) and Grp94. Hsp90a, Hsp90b are essential for the viability of eukaryotic cells. They are rather abundant constitutively, make up 1–2% of cytosolic proteins, and their expression level can be further stimulated by stress. Whereas Hsp90b is more or less constitutively and ubiquitously expressed, Hsp90a is heat-inducible and has a more tissue-specific expression. Hsp90 seems to be more selective than other chaperones to interact with proteins, and regulates the biological function of many proteins that are signaling factors. Generally, Hsp90 does not function as an individual chaperone, but as a component of larger complexes. In fact, steroid receptors are able to form heterocomplexes with the chaperones Hsp90 and Hsp70, and the co-chaperone p23 (Grad and Picard, 2007; Galigniani et al., 2010). On the other hand, Sreedhar et al. (2004) demonstrated that Hsp90 seems to have different molecular partners depending on the apoptotic stimuli, the effect of the protein being predominantly antiapoptotic. Other functions of Hsp90 have been proposed to be related to chromatin remodeling and epigenetic programming related to cancer (Ruden et al., 2005), in innate immunity (Kadota et al., 2010) and in neurodegenerative diseases (Luo et al., 2010; Salminen et al., 2011).

The best known members of the HSP70 family are: the 73-kDa protein which is constantly produced, hence the term "constitutive" (Hsc70), the 72-kDa protein which is highly inducible and its synthesis is increased in response to multiple stressors (Hsp70)

(Kregel, 2002), the mitochondrial protein which has a molecular weight of 75 kDa (Hsp75) and the endoplasmic reticulum protein which has a molecular weight of 78 kDa (Grp78). Chaperokine is a term recently used to describe the peculiar function of extracellular Hsp70 (eHsp72) as both chaperone and cytokine (Asea 2003, 2005). The consequence of binding and signaling is the stimulation of a potent and long-lasting immune response.

In mammals, the HSP60 (chaperonin) family consists of mitochondrial Hsp60 (mt-Hsp60) and cytosolic Hsp60. HSP60 members are believed to be predominantly mitochondrial, although there are reports in extracellular compartments (Gupta 1990; Retzlaff et al. 1994; Sarkar et al. 2006). As a molecular chaperone, Hsp60 helps in the folding of nascent polypeptides and in the transport of proteins from the cytoplasm to organelles (Fink, 1999). In addition to their typical chaperon function, these proteins are also implicated in other diverse activities such as amino acid transport, signal transduction, peptide presentation, regulation of the immune system, and apoptosis (Ikawa and Weinberg 1992; Jones et al. 1994; Wells et al. 1997; Woodlock et al. 1997; Sarkar et al. 2006). mt-Hsp60 exists in a dynamic equilibrium among monomers, heptamers, and tetradecamers (Fink, 1999). It dissociates into monomers at low concentrations and assembles into tetradecamers in the presence of ATP and mt-Hsp10, the cofactor of mt-Hsp60 (Levy-Rimler G. et al., 2001). The cytosolic Hsp60 forms hetero-oligomeric ring structures and functions in the cytosol to fold cytoskeletal proteins such as actin and tubulin (Llorca et al., 2000). Both Hsp60 and Hsp10 have been demonstrated to have a proapoptotic role during apoptosis, since the activation of caspase-3 occurs simultaneously with Hsp60 and Hsp10 release from mitochondria (Garrido et al., 2001). In contrast, increased expression of Hsp60 obtained by stable transfection, which has been shown to have a protective effect against apoptosis, is unlikely to occur under physiological conditions, Hsp60 being mainly constitutive. Thus, cytosolic Hsp60 prevents translocation of the pro-apoptotic protein Bax into mitochondria and thus promotes cell survival but also promotes maturation of procaspase-3, essential for caspase-mediated cell death (Arya et al., 2007). The presence of secreted Hsp60 in the bloodstream is consistent with several inflammatory conditions (Henderson and Pockley, 2010) and it is clear that extracellular Hsp60 is a link between body tissues and the immune system (Quintana and Cohen, 2011).

The defining feature of the Hsp40 chaperone family is a ~70-amino-acid-residue signature, termed the J domain, which is necessary for orchestrating interactions with its Hsp70 chaperone partner(s). J-domain proteins play important regulatory roles as co-chaperones, recruiting Hsp70 partners and accelerating the ATP-hydrolysis step of the chaperone cycle. Because the intrinsic ATPase activity of all Hsp70s is extremely weak, co-chaperones such as Hsp40 are needed for stimulation of ATP hydrolysis. Hsp40 is thought to change the conformation of Hsp70 to a form that displays a higher affinity for its various substrates following ATP hydrolysis. Because ATP binding and hydrolysis perform the mechanical work necessary to regulate conformational changes in the Hsp70 substrate-binding domain, it is clear that co-chaperones such as Hsp40 are key regulators of the ubiquitous Hsp70 chaperone. J-domain proteins participate in complex biological processes, such as cell cycle control by DNA tumor viruses, regulation of protein kinases and exocytosis (Kelley et al., 1998; Gotoh et al., 2004).

Hsp27 belongs to the subfamily of small HSPs, and is expressed in many cell types and tissues, at specific stages of development and differentiation (Garrido et al., 1999). Hsp27 is an ATP-independent powerful chaperone and recent evidence has shown that it regulates apoptosis through the interaction with key components of the apoptotic signaling pathway and that, in particular, it is involved in caspase activation and apoptosis (Concannon et al., 2003). Overexpression of Hsp27 prevents the cell death triggered by various stimuli including hyperthermia, oxidative stress, and cytotoxic drugs. This overexpression provides an example of how pro-apoptotic stimuli, delivered below a threshold level, can elicit protective responses (Garrido et al., 2003).

2. Involvement of HSPs in Stress Conditions

Several studies have demonstrated that HSPs are also expressed under physiological conditions and that they play an important role in normal cell functions (Lindquist and Craig, 1988). In fact, many members of these HSPs are present constitutively (heat shock cognate) in cells while some are expressed only after stress (Beere, 2004). Numerous studies indicate that, in stress conditions, these proteins are involved in the regulation of cell growth, transformation and death pathways (Helmbrecht et al., 2000). Moreover, their expression would be part of the functional response to the changes in hormones and neurotransmitters induced by stress and there are reports indicating that HSPs can control hormonal functions and vice versa.

Stress-inducing agents often affect the redox state and hydration of the cell, which, in turn, cause increased levels of misfolded proteins that may be deleterious by virtue of their altered biological activities. The cellular response to stress is represented at the molecular level by the induced synthesis of HSPs. Exposure of cells to acute and chronic stress shifts the protein-folding equilibrium, such that molecular chaperones are directed toward the capture of folding intermediates to prevent misfolding and aggregation and to facilitate refolding or degradation (Ellis, 1993; Welch, 1993).

Recent studies suggest that HSPs are upregulated in both the central and peripheral nervous systems after injury (Kim et al., 2001), and that they play an important role in neuronal survival (Dodge et al., 2006). Hsp27, in particular, has been extensively studied in the peripheral nervous system, where it inhibits apoptosis after injury (Klass et al., 2008).

Further studies have indicated that the potent cytoprotective and folding properties of HSPs are co-opted during malignant progression when HSPs become expressed at high levels to facilitate tumor cell growth and survival (Ciocca and Calderwood, 2005). However, there is no convincing evidence to indicate that increased levels of HSPs are essential for the inactivation of tumor suppressor molecules that is necessary for transformation (Calderwood et al., 2006).

A growing body of evidence suggests that HSPs are also closely involved in a number of crucial processes in tumor development such as the regulation of cell cycle progression (Helmbrecht et al., 2000), control of apoptotic pathways (Didelot et al., 2006; Garrido et al., 2006; Schmitt et al., 2007) and immune surveillance against cancer (Li, 2001). Indeed, studies are underway to determine whether these proteins could be used as diagnostic and/or prognostic markers or represent new target for therapy (Romanucci et al., 2008).

3. Involvement of HSPs in the Ovarian Function

HSPs participate in two key processes of ovarian physiology: proliferation/apoptotic balance and the action of steroids hormones. It is well known that adequate conditions promote reproduction via hormonal stimulators while adverse conditions (for example, inadequate temperature, nutritional deficiencies, and infections) induce stress and suppress reproduction through stress-related substances. These anomalies induce changes in the expression of numerous genes, including genes encoding HSPs (Sirotkin and Bauer, 2011).

It has been found that some stress proteins are highly expressed in the female reproductive organs (Ciocca et al., 1996). The ovarian cycle is associated with changes in HSPs expression and these proteins are very important in ovarian physiology, particularly in follicular development through modulation of steroidal hormones functions as mentioned above. It is well known that Hsp90 and Hsp70 modulate the function of sex steroid receptor proteins (Bagchi et al., 1991; Pratt and Toft, 1997; Ravagnan et al., 2001). Based on *in vitro* experiments, dissociation of Hsp70 and Hsp90 from the receptor protein before binding of the steroid-receptor complex to DNA is essential for subsequent transcriptional activation (Bagchi et al., 1991).

HSPs are also involved in cell survival and death mechanisms. Evidence has accumulated that cell cycle components, regulatory proteins and members of the mitogenic signal cascade associate with chaperones and stress proteins for different periods of time (Helmbrecht et al., 2000). The HSP27, HSP70 and HSP90 families have been implicated in the protection against apoptosis induced by a variety of stimuli such as heat shock, nutrient withdrawal, reactive oxygen species, endoplasmic reticulum stresses, proteasome inhibition, and cytoskeletal perturbation (Beere, 2005). In relation to apoptotic cell death, which mediates the elimination of damaged or unwanted cells, the underlying ability of HSPs to maintain cell survival correlates with the inhibition of the mechanism of caspase activation (Garrido et al., 1999; Beere et al., 2000; Mosser et al., 2000; Beere, 2004).

In regard to cell growth, the prevalence of chaperones/HSP is increased in proliferating cells as compared to that in cells in the stationary state or differentiated cells (Wu et al., 1986; Hensold and Houseman, 1988; Sainis et al., 1994; Helmbrecht and Rensing, 1999). Several studies have evaluated the amounts of HSPs in the ovaries of laboratory animals (Khanna et al., 1995; Ohsako et al., 1995; Paranko et al., 1996; Maizels et al., 1998; Salvetti et al., 2009), bovine oocytes (Wrenzycki et al., 2001), bovine follicular fluid (Driancourt et al., 1999; Driancourt, 2001; Maniwa et al., 2005) and human ovarian cells (Langdon et al., 1995; Kim et al., 1996) as well as the constitutive amounts of these proteins in ovaries of adult cattle (Velázquez et al., 2010).

Previous observations of aberrant HSPs gene expression in cells of the ovarian cystic follicles suggest that these proteins could be associated with the intra-ovarian component of cystic ovarian disease (COD) pathogenesis, due to a link with intracellular apoptosis signaling pathways. Because proliferation and apoptosis are part of normal follicular development and atresia, some authors have assumed that disorders in the balance between cell proliferation and apoptosis may be associated with cystogenesis (Isobe and Yoshimura, 2007; Salvetti et al., 2010).

Several studies about Hsp10, Hsp27, Hsp40, Hsp60, Hsp70 and Hsp90α/β expression have allowed the localization and quantification of their mRNAs and expressed proteins in each ovarian component, and thus the determination of any associations with healthy and cystic follicles (Velázquez et al., 2010; 2011).

4. HSPs and Ovarian Disorders

Cystic ovarian disease (COD) and Polycystic ovarian syndrome (PCOS) are reproductive disorders that affect many species of zootechnical interest (COD) and humans (PCOS) (Jakimiuk et al., 2002; Silvia et al., 2002; Vanholder et al., 2006). The heterogeneity of the disease is reflected in many animal models of polycystic ovaries (PCO) (Mahesh et al., 1987; Mahajan, 1988; Salvetti et al., 2004; Baravalle et al., 2006; Ortega et al., 2009).

In the bovine livestock, COD is an important cause of infertility characterized by anovulation, anestrus, and the persistence of follicles with a diameter larger than that of the ovulatory follicle (Silvia et al., 2002). COD has been defined as the presence of one or more follicular structures in the ovary/ovaries, with a diameter of at least 20 mm, which persist for more than 10 days in the absence of luteal tissue, interrupting the normal reproductive cycle (Silvia et al., 2002; Peter, 2004; Vanholder et al., 2006). Many factors such as stress, nutritional management and infectious disease can coexist in animals with COD; however, the primary cause has not yet been elucidated. Although it is widely accepted that dysfunction of the hypothalamic-pituitary-gonadal axis is an important etiologic factor of COD, delay of follicle regression after ovulation failure is an alternative cause of cysts (Salvetti et al., 2010). Alterations in the ovarian micro-environment of females that present follicular cysts could alter the normal processes of proliferation and programmed cell death in ovarian cells.

In a recent work levels of HSPs in relation to follicular development were evaluated in ovaries of healthy cattle and those of cows with COD. The expression of Hsp10 mRNA showed lower levels of mRNA amount in granulosa and theca layers of cystic follicles than in those of healthy follicles. In fact, the increased level of Hsp10 mRNA observed in normal follicles might be associated mainly with cell survival (Velázquez et al., 2011). In this sense, Ling et al. (2011) found that Hsp10 protein may be involved in the regulation of several biological functions, including protection of *in vitro* cultured healthy follicular cells against apoptosis. In the same way, ovaries of women with PCOS have been reported to express lower levels of Hsp10 than healthy ovaries (Ma et al., 2007) and these results are similar to those we found in ovaries from cows with COD (Velázquez et al., 2011). Taking together, these findings allow assuming that apoptosis mechanisms would be controlled factors other than Hsp10, such as Bcl-2, Hsp27 and Hsp70 (Cascales-Angosto, 2003).

There are numerous reports that sustain a role for Hsp27 in PCOS pathogenesis and this protein also seems to be an important regulator of signal transduction pathways, differentiation and apoptosis (Kostenko and Moens, 2009; Rayner et al., 2010). In this sense, both Hsp27 and Hsp70 can modulate Bid-dependent apoptosis and Hsp27 specifically interacts with cytochrome c in the cytosol, an interaction with functional consequences since it prevents the formation of the apoptosome (Garrido et al., 2001; Beere, 2005).

In human tissues, Hsp27 is highly expressed in estrogen target organs of the female reproductive tract (Ciocca et al., 1996). This protein is under hormonal modulation and significant changes in Hsp27 localization and content have been observed in the human endometrium during the different phases of the menstrual cycle (Ciocca et al., 1993).

Recent studies in ovaries from cattle have shown that Hsp27 is constant in the granulosa of secondary, tertiary and cystic follicles, but varies in theca cells with follicular development (Figure 1) (Velázquez et al., 2010, 2011). In addition, significantly lower expression has been observed in the granulosa and theca cells of atretic follicles. Coincidentally, a lower index of apoptosis is observed in cystic ovarian follicles in cows (Isobe and Yoshimura, 2007; Ortega et al., 2008; Salvetti et al., 2010). It has been suggested that a disorder of cell proliferation and apoptosis is associated with the occurrence of follicular cysts. In the same way, Hsp27 mRNA expression is significantly lower in granulosa cells of cystic and large follicles than in small and medium follicles. In theca cells, the highest expression of Hsp27 mRNA has been detected in healthy medium follicles (Velázquez et al., 2011). These findings are consistent with the notion that Hsp27 prevents cell apoptosis during normal folliculogenesis.

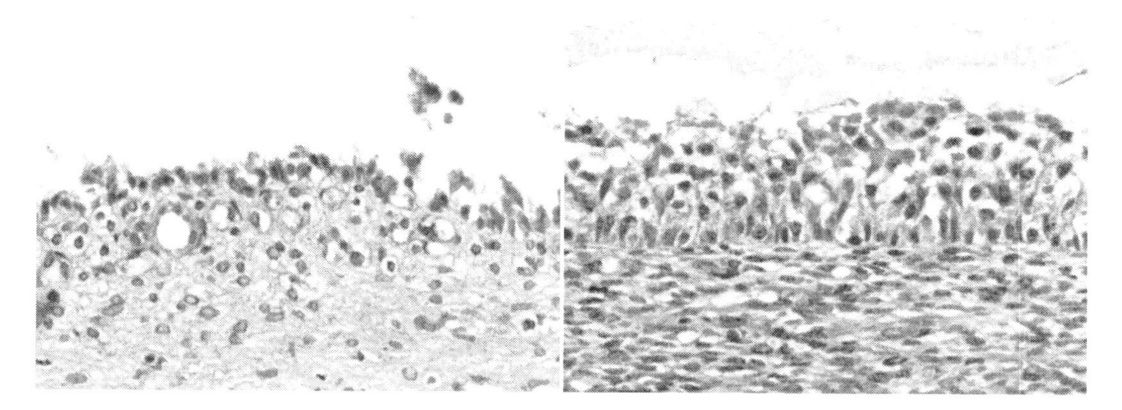

Figure 1. Immunohistochemical staining with Hsp27 antibody of tertiary (right) and cystic follicles (left) of ovaries from healthy cows and with COD. X40.

It should be emphasized that Hsp40 protein functions as a co-chaperone of mammalian Hsp70. There are few data reported regarding Hsp40 expression in ovaries. Analyses performed in tissues from women with PCOS have demonstrated that Hsp40 is differentially expressed and have identified as a potential susceptibility marker in PCOS (Jansen et al., 2004; Jones et al., 2011).

Recent data have also shown that Hsp70 and its co-chaperone Hsp40 have a cooperative role in the inhibition of Bax translocation from the cytosol to mitochondria to prevent apoptosis (Gotoh et al., 2004). These observations indicate that, whereas HSPs can function alone to inhibit apoptosis, cooperative interaction with their designated co-chaperone molecules is likely to enhance their anti-apoptotic activities (Beere, 2004).

In the ovary, the mRNA expression of Hsp40 in granulosa and theca cell layers of cysts is lower than that in normal follicles. Similar results have been observed with Hsp70 mRNA in theca cells (Velázquez et al., 2011).

Hsp60 is present in human follicular fluid (Neuer et al., 1997) and is related to fertility (Neuer et al., 2000). Greater amounts of Hsp60 have been found in the oocytes of single-layered primordial follicles and in those of the growing preantral follicles of the rat than in the oocytes of antral follicles. In addition, preovulatory oocytes have been found to remain distinctly immunoreactive (Paranko et al., 1996). This suggests that these active cells contain a key element required for proper refolding of imported mitochondrial proteins (Paranko et al., 1996). Also, like other HSPs, Hsp60 can stabilize steroid receptors (Edwards et al., 1992); this stabilization deprives them of their capacity to enter the nucleus and initiate their nuclear function (Kiang and Tsokos, 1998; Neuer et al., 2000).

In a recent work (Velázquez et al., 2010) we observed intense immunostaining of Hsp60 in the granulosa of atretic follicles, with lower staining in primary and secondary follicles in the ovary of cows and greater amounts of staining in the cells of the theca interna in tertiary and cystic follicles (Figure 2). We also found higher expression of Hsp60 mRNA in granulosa cells of medium follicles than in those of other healthy follicles and cysts, and that in theca cells the highest expression was in cystic follicles. Similarly, we detected more intense immunohistochemical staining of Hsp60 in the theca cells of cystic follicles than in those of healthy follicles. Because mitochondrial and nuclear HSPs may play a role in the maintenance of metabolic activity and survival of the oocyte (Ohsako et al., 1995) this may be a reason for the higher level of HSP60 observed in the granulosa cell layer of healthy medium follicles (Velázquez et al., 2010).

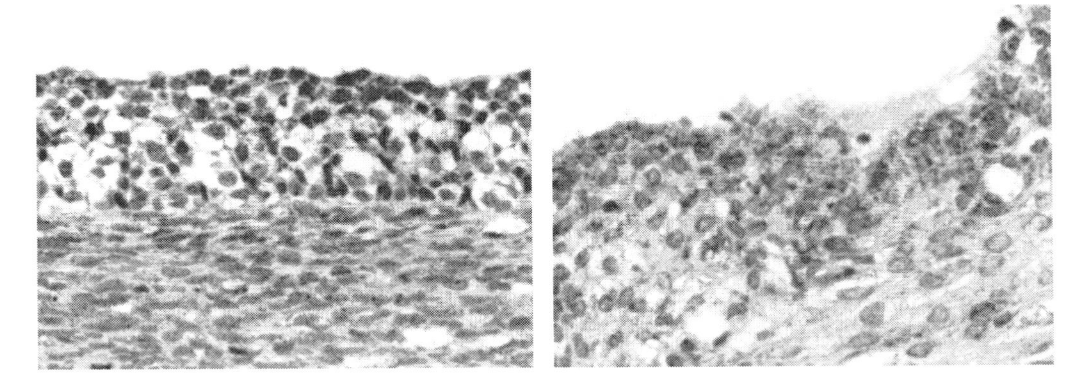

Figure 2. Immunohistochemical staining with Hsp60 antibody of tertiary (left) and cystic follicles (right) of ovaries from healthy cows and with COD. X40.

It has become widely recognized that Hsp70 is able to inhibit apoptosis and that this ability may contribute to its protective effect against cell death (Beere, 2004, 2005). As mentioned above, Isobe and Yoshimura (2007) reported that in the theca interna of ovarian follicles of cattle, the frequency of apoptosis is greater in early cystic follicles than in late cystic follicles. These authors concluded that the decrease in the apoptosis cell rate may be responsible for the delay in follicular regression, and that control of apoptosis may be essential for reducing the incidence of cystic follicles. This observation is consistent with our results and other previous research, in which a decreased apoptotic cell death was associated with increased Hsp70 in cystic follicles (Maniwa et al., 2005; Salvetti et al., 2009, 2010; Velázquez et al., 2010).

The expression of Hsp70 is probably associated with aberrant expression of steroid hormone receptors in follicles of cows with COD (Espey, 1994; Garrido et al., 2003; Salvetti et al., 2007, 2009). Recently, we detected lower expression of Hsp70 mRNA in theca cells of cystic follicles than those of healthy follicles, and that granulosa cells of small and medium follicles showed higher expression levels than large and cystic follicles (Velázquez et al., 2011). In another previous work, we found a decrease in the expression of estrogen receptor β (ERβ) in growing, atretic, and cystic follicles in ovaries from cows with COD compared with tertiary follicles from control animals. We also found a high expression of ERα in cystic follicles in relation to other follicular categories (Salvetti et al., 2007).

Hsp90 has also been observed to have a staining distribution and relative amounts similar to those of Hsp70 in the ovary. Hsp90 increases during follicular development, in agreement with previous findings (Driancourt et al., 2001) where Hsp90 is also present in cattle ovaries, with follicular size- and atresia-related changes in its concentration. In addition, granulosa cells have been found to be more active producers of Hsp90 than theca cells. Therefore, Hsp90 may be involved in follicular maturation (Velázquez et al., 2010).

Hsp90 is a molecular chaperone protein that regulates signal transduction by nuclear receptors (NR) and protein kinases. This molecular chaperone associates with the unliganded form of estrogen receptors, as well as with the androgen receptor, the progesterone receptor, the glucocorticoid receptor, and the mineralocorticoid receptor. Ligand binding to each NR induces a conformational change within the receptor, causing its dissociation from Hsp90, leading to receptor dimerization, interaction with cofactors, DNA binding, and target gene activation. Thus, molecular chaperoning is an essential initial step in the tightly regulated process of ligand-dependent transcriptional control of ER and could thus be involved in the regulation of steroidogenesis and follicular development (Johnson and Toft, 1994; Fang et al., 1996; Powell et al., 2010).

In rodents, Hsp90 may be present in germinal (Curci et al., 1991) and somatic cells (Ben-Ze'ev and Amsterdam, 1989) of the ovary. In previous studies carried out in rats, researchers noted that the subchronic treatment with a low dose of atrazine kept the Hsp90 expression in the oocytes, while increasing its expression in the granulosa cells. Hsp90 is normally accumulated in the oocytes and may be essential for female mammalian premeiotic germ cells (Ohsako et al., 1995). Hsp90 overexpression in granulosa cells is probably triggered by atrazine exposure since, in normal conditions, ovarian somatic cells, including follicular granulosa cells, show only faint reactivity (Ohsako et al., 1995). It is probable that the synthesis of stress-induced Hsp90 in the follicles allows cells to adapt and to avoid the cell death induced by atrazine in the granulosa cells as well as in the oocyte. In this situation, Hsp90 acts as an antiapoptotic protein (Garrido et al., 2001). This adaptation occurs because Hsp90 may act as a chaperone, taking into account the large quantity of unfolded or denatured proteins, which are most likely present in these stressed cells. However, when the atrazine concentrations are higher than the cell ability to adapt itself, cell death becomes a predominant phenomenon (Juliani et al., 2008).

In a recent study we observed some differences between Hsp90b and Hsp90a expression in tissues from cattle with COD (Velázquez et al., 2011). Hsp90b mRNA expression in theca and granulosa cells was lower in cystic follicles than healthy follicles. In contrast, we detected a high protein level of Hsp90 in cysts compared with the remaining follicles. However, the high expression of Hsp90a mRNA found in follicular cysts was correlated with the protein level. Differences between mRNA and protein expression of Hsp90 may be a part of the cellular deregulation that occurs during the development of COD. Generally, the relative levels of HSP mRNAs observed are consistent with the levels of HSP protein reported previously (Velázquez et al., 2010).

Conclusion

Results obtained by multiple authors, it can be concluded that abnormal amounts of HSP are related to ovarian diseases, which suggests that their expression levels could be closely associated with a regulatory mechanism during follicular development. Moreover, the profiles are consistent with altered patterns of other factors involved in the persistence and arrest of follicular structures in ovarian cysts.

References

Arya, R; Mallik, M & Lakhotia, SC (2007) Heat shock genes – integrating cell survival and death. *Journal of Biosciences* 32:595–610.

Asea, A (2003) Chaperokine-induced signal transduction pathways. *Exercise Immunology Review* 9:25–33.

Asea, A (2005) Stress proteins and initiation of immune response: chaperokine activity of hsp72. *Exercise Immunology Review* 11:34–45.

Bagchi, MK; Tsai, SY; Tsai, MJ & O'Malley, BW (1991) Progesterone enhances target gene transcription by receptor free of heat shock proteins hsp90, hsp56, and hsp70. *Molecular and Cellular Biology* 11:4998-5004.

Baravalle, C; Salvetti, NR; Mira, GA; Pezzone, N & Ortega HH (2006) Microscopic characterization of follicular structures in letrozole-induced polycystic ovarian syndrome in the rat. *Archives of Medical Research* 37:830-839.

Beere, HM (2004) 'The stress of dying': the role of heat shock proteins in the regulation of apoptosis. *Journal of Cell Science* 117: 2641-2651.

Beere, HM (2005) Death versus survival: functional interaction between the apoptotic and stress-inducible heat shock protein pathways. *Journal of Clinical Investigation* 115: 2633–2639.

Beere HM; Wolf, BB; Cain, K; Mosser, DD; Mahboubi, A; Kuwana, T; Tailor, P; Morimoto, RI; Cohen, GM & Green, DR (2000) Heat-shock protein 70 inhibits apoptosis by preventing recruitment of procaspase-9 to the Apaf-1 apoptosome. *Nature Cell Biology* 2:469-475.

Ben-Ze'ev, A & Amsterdam, A (1989) A regulation of heat shock protein synthesis by gonadotropins in cultured granulosa cells. *Endocrinology* 124:2584–2594.

Calderwood, SK; Khaleque, MA; Sawyer, DB & Ciocca, DR (2006) Heat shock proteins in cancer: chaperones of tumorigenesis. *Trends in Biochemical Sciences* 31: 164-172.

Cascales Angosto M (2003) Bases moleculares de la apoptosis. *Anales de la Real Academia Nacional de Farmacia* 69:36-64.

Ciocca, DR; Oesterreich, S; Chamness, GC; McGuire, WL & Fuqua, SAW (1993) Heat shock protein 27,000 (hsp27): biological and clinical implications. *Journal of the National Cancer Institute* 85:1558-1570.

Ciocca, DR; Stati, AO; Fanelli, MA & Gaestel, M (1996) Expression of heat shock protein 25,000 in rat uterus during pregnancy and pseudopregnancy. *Biology of Reproduction* 54:1326-1335.

Ciocca, DR & Calderwood, SK (2005) Heat shock proteins in cancer: diagnostic, prognostic, predictive, and treatment implications. *Cell Stress & Chaperones* 10:86–103.

Concannon, CG; Gorman, AM & Samali, A (2003) On the role of Hsp27 in regulating apoptosis. *Apoptosis* 8:61–70.

Curci, A; Bevilacqua, A; Fiorenza, MT & Mangia, F (1991) Developmental regulation of heat-shock response in mouse oogenesis: identification of differentially responsive oocyte classes during Graafian follicle development. *Developmental Biology* 144:362–368.

Didelot, C; Schmitt, E; Brunet, M; Maingret, L; Parcellier, A & Garrido, C (2006) Heat shock proteins: endogenous modulators of apoptotic cell death. *Handbook of Experimental Pharmacology* 172:171–198.

Dodge, ME; Wang, J; Guy, C; Rankin, S; Rahimtula, M & Mearow, KM (2006) Stress-induced heat shock protein 27 expression and its role in dorsal root ganglion neuronal survival. *Brain Research* 1068:34-48.

Driancourt, MA (2001) Regulation of ovarian follicular dynamics in farm animals. Implications for manipulation of reproduction. *Theriogenology* 55:1211–1239.

Driancourt, MA; Guet, P; Reynaud, K; Chadli, A & Catelli, MG (1999) Presence of an aromatase inhibitor, possibly heat shock protein 90, in dominant follicles of cattle. *Journal of Reproduction and Fertility* 115:45-58.

Edwards, DP; Estes, PA; Fadok, VA; Bona, BJ; Onate, S; Nordeen, SK & Welch, WJ (1992) Heat shock alters the composition of heteromeric steroid receptor complexes and enhances receptor activity in vivo. *Biochemistry* 31:2482–2491.

Ellis, RJ (1993) The general concept of molecular chaperones. *Philosophical Transactions of the Royal Society B: Biological Sciences* 339:257–261.

Espey, LL (1994) Current status of the hypothesis that mammalian ovulation is comparable to an inflammatory reaction. *Biology of Reproduction* 50:233-238.

Fang, Y; Fliss, AE; Robins, DM & Caplan, AJ (1996) Hsp90 regulates androgen receptor hormone binding affinity in vivo. *Journal of Biological Chemistry* 271: 28697-702.

Fink, AL (1999) Chaperone-mediated protein folding. Physiological Reviews 79:425–449.

Galigniana, MD; Echeverría, PC; Erlejman, A & Piwien-Pilipuk, G (2010) Role of molecular chaperones and TPR-domain proteins in the cytoplasmic transport of steroid receptors and their passage through the nuclear pore. *Nucleus* 1:299-308.

Garrido, C; Bruey, JM; Fromentin, A; Hammann, A; Arrigo, AP & Solary, E (1999) HSP27 inhibits cytochrome c-dependent activation of procaspase-9. *FASEB Journal* 13: 2061–2070.

Garrido, C; Gurbuxani, S; Ravagnan, L & Kroemer, G (2001) Heat shock proteins: endogenous modulators of apoptotic cell death. *Biochemical and Biophysical Research Communications* 286:433–442.

Garrido, C; Schmitt, E; Candé, C; Vahsen, N; Parcellier, A & Kroemer, G (2003) HSP27 and HSP70: potentially oncogenic apoptosis inhibitors. *Cell Cycle* 2:579-584.

Garrido, C; Brunet, M; Didelot, C; Zermati, Y; Schmitt, E & Kroemer, G (2006) Heat shock proteins 27 and 70. Anti-apoptotic proteins with tumorigenic properties. *Cell Cycle* 5:2592–2601.

Gotoh, T; Terada, K; Oyadomari, S & Mori, M (2004) hsp70-DnaJ chaperone pair prevents nitric oxide- and CHOP-induced apoptosis by inhibiting translocation of Bax to mitochondria. *Cell Death Differentiation* 11:390-402.

Grad, I & Picard, D (2007) The glucocorticoid responses are shaped by molecular chaperones. *Molecular and Cellular Endocrinology* 275:2-12.

Gupta, RS (1990) Microtubules, mitochondria, and molecular chaperones: a new hypothesis for in vivo assembly of microtubules. *Biochemistry and Cell Biology* 68:1352–1363.

Hatayama, T; Yamagishi, N; Minobe, E & Sakai, K (2001) Role of hsp105 in protection against stress-induced apoptosis in neuronal PC12 cells. *Biochemical and Biophysical Research Communications* 288:528-534.

Helmbrecht, K & Rensing, L (1999) Different constitutive heat shock protein 70 expression during proliferation and differentiation of rat C6 glioma cells. *Neurochemical Research* 24:1293-1299.

Helmbrecht, K; Zeise, E & Rensing, L (2000) Chaperones in cell cycle regulation and mitogenic signal transduction: a review. *Cell Proliferation* 33:341–365.

Henderson, B & Pockley, A. (2010) Molecular chaperones and protein-folding catalysts as intercellular signaling regulators in immunity and inflammation. *Journal of Leukocyte Biology* 88:445-462.

Hensold, JO & Housman, DE (1988) Decreased expression of the stress protein HSP70 is an early event in murine erythroleukemic cell differentiation. *Molecullar and Cellular Biology* 8:2219-2223.

Ikawa, S & Weinberg, RA (1992) An interaction between p21ras and heat shock protein hsp60, a chaperonin. *Proceedings of the National Academy of Sciences of United States of America* 89:2012–2016.

Isobe, N & Yoshimura, Y (2007) Deficient proliferation and apoptosis in the granulosa and theca interna cells of the bovine cystic follicle. *Journal of Reproduction and Development* 53:1119–1124.

Jakimiuk, AJ; Weitsman, SR; Yen, HW; Bogusiewicz, M & Magoffin, DA (2002) Estrogen receptor a and b expression in theca and granulosa cells from women with polycystic ovary syndrome. *Journal of Clinical Endocrinology and Metabolism* 87: 5532–5538.

Jansen, E; Laven, JS; Dommerholt, HB; Polman, J; van Rijt, C; van den Hurk, C; Westland, J; Mosselman, S & Fauser, BC (2004) Abnormal gene expression profiles in human ovaries from polycystic ovary syndrome patients. *Molecular Endocrinology* 18: 3050-3063.

Johnson, JL & Toft DO (1994) A novel chaperone complex for steroid receptors involving heat shock proteins, immunophilins, and p23. *Journal of Biological Chemistry* 269:24989-24993.

Jones, M; Gupta, RS; Englesberg, E (1994) Enhancement in amount of P1 (hsp60) in mutants of Chinese hamster ovary (CHO-K1) cellos exhibiting increases in the A system of amino acid transport. *Proceedings of the National Academy of Sciences of United States of America* 91:858–862.

Jones, MR; Chua, A; Chen, Y-DI; Li, X; Krauss, RM; Rotter, JI; Legro, RS; Azziz, R & Goodarzi, MO (2011) Harnessing Expression Data to Identify Novel Candidate Genes in Polycystic Ovary Syndrome. *PLoS ONE* 6:e20120.

Juliani, CC; Silva-Zacarin, EC; Santos, DC & Boer, PA (2008) Effects of atrazine on female Wistar rats: morphological alterations in ovarian follicles and immunocytochemical labeling of 90 kDa heat shock protein. *Micron* 39:607-616.

Kadota, Y; Shirasu, K & Guerois, R (2010) NLR sensors meet at the SGT1-HSP90 crossroad. *Trends in Biochemical Science* 35:199-207.

Kelley, WL (1998) The J-domain family and the recruitment of chaperone power. *Trends in Biochemical Science* 23.

Khanna, A; Aten, RF & Behrman, HR (1995) Heat shock protein-70 induction mediates luteal regression in the rat. *Molecular Endocrinology* 9:1431–1440.

Kiang, JG & Tsokos, GC (1998) Heat shock protein 70 kDa: molecular biology, biochemistry, and physiology. *Pharmacology and Therapeutics* 80: 183-201.

Kim, AH; Khanna, A; Aten, RF; Olive, DL & Behrman, HR (1996) Cytokine induction of heat shock protein in human granulosa-luteal cells. *Molecular Human Reproduction* 2:549–554.

Kim, DS; Lee, SJ; Park, SY; Yoo, HJ; Kim, SH; Kim, KJ & Cho HJ (2001) Differentially expressed genes in rat dorsal root ganglia following peripheral nerve injury. *Neuroreport* 12:3401-3405.

Klass, MG; Gavrikov, V; Krishnamoorthy, M & Csete, M (2008) Heat shock proteins, endothelin, and peripheral neuronal injury. *Neuroscience Letters* 433:188-193.

Kostenko, S & Moens, U (2009) Heat shock protein 27 phosphorylation: kinases, phosphatases, functions and pathology. *Cell and Molecular Life Sciences* 66:3289-3307.

Kregel, KC (2002) Molecular Biology of Thermoregulation. Invited review: Heat shock proteins: modifying factors in physiological stress responses and acquired thermotolerance. *Journal of Applied Physiology* 92:2177–2186.

Langdon, SP; Rabiasz, GJ; Hirst, GL; King, RJ; Hawkins, RA; Smyth, JF & Miller, WR (1995) Expression of the heat shock protein HSP27 in human ovarian cancer. *Clinical Cancer Research* 1:1603–1609.

Levy-Rimler, G; Viitanen, P; Weiss, C; Sharkia, R; Greenberg, A; Niv, A; Lustig, A; Delarea, Y & Azem, A (2001) The effect of nucleotides and mitochondrial chaperonin 10 on the structure and chaperone activity of mitochondrial chaperonin 60. *European Journal of Biochemistry* 268: 3465-3472.

Li, Z (2001) The roles of heat shock proteins in tumour immunity. In: Giaccone G, Schilsky R, Sondel P (eds) *Cancer chemotherapy and biological response modifiers.* Elsevier, New York, pp 371–382.

Lindquist, S (1986) The heat-shock response. *Annual Review of Biochemistry* 55:1151-1191.

Lindquist, S & Craig, EA (1988) The heat shock proteins. *Annual Review of Genetics.* 22:631– 637.

Ling, J; Zhao, K; Cui, YG; Li, Y; Wang, X; Li, M; Xue, K; Ma, X & Liu, JY (2011) Heat shock protein 10 regulated apoptosis of mouse ovarian granulosa cells. *Gynecological Endocrinology* 27:63-71.

Llorca, O; Martin-Benito, J; Grantham, J; Ritco-Vonsovici, M; Willison, KR; Carrascosa, JL & Valpuesta, JM (2001b) The 'sequential allosteric ring' mechanism in the eukaryotic chaperonin-assisted folding of actin and tubulin. *EMBO Journal* 20: 4065–4075.

Luo, W; Sun, W; Taldone, T; Rodina, A & Chiosis, G (2010) Heat shock protein 90 in neurodegenerative diseases. *Molecular Neurodegeneration* 5:24.

Ma, X; Fan, L; Meng, Y; Hou, Z; Mao, YD; Wang, W; Ding, W & Liu, JY (2007) Proteomic analysis of human ovaries from normal and polycystic ovarian syndrome. *Molecular Human Reproduction* 13:527–535.

Mahajan, DK (1988) Polycystic ovarian disease: animal models. *Endocrinology and Metabolism Clinics of North America* 17:705–732.

Mahesh, VB; Mills, TM; Bagnell, CA & Conway, BA (1987) Animal models for study of polycystic ovaries and ovarian atresia. *Advances in Experimental Medicine and Biology* 219:237–257.

Maizels, ET; Cottom, J; Jones, JC & Hunzicker-Dunn, M (1998) Follicle stimulating hormone (FSH) activates the p38 mitogen-activated protein kinase pathway, inducing small heat shock protein phosphorylation and cell rounding in immature rat ovarian granulosa cells. *Endocrinology* 139:3353-3356.

Maniwa, J; Izumi, S; Isobe, N & Terada, T (2005) Studies on substantially increased proteins in follicular fluid of bovine ovarian follicular cysts using 2-D PAGE and MALDI-TOF MS. *Reproductive Biology and Endocrinology* 3, 23.

Morimoto, RI (1993) Cells in stress: transcriptional activation of heat shock genes. *Science* 259:1409-1410.

Mosser, DD; Caron, AW; Bourget, L; Meriin, AB; Sherman, MY; Morimoto, RI & Massie, B (2000) The chaperone function of hsp70 is required for protection against stress-induced apoptosis. *Molecular and Cellular Biology* 19:7146–7159.

Neuer, A; Lam, KN; Tiller, FW; Kiesel, L & Witkin, SS (1997) Humoral immune response to membrane components of Chlamydia trachomatis and expression of human 60 kDa heat shock protein in follicular fluid of in-vitro fertilization patients. *Human Reproduction* 12:925–929.

Neuer, A; Spandorfer, SD; Giraldo, P; Dieterle, S; Rosenwaks, Z & Witkin, SS (2000) The role of heat shock proteins in reproduction. *Human Reproduction Update* 6:149–159.

Ohsako, S; Bunick, D & Hayashi, Y (1995) Immunocytochemical observation of the 90 kD heat shock protein (HSP90): high expression in primordial and pre-meiotic germ cells of male and female rat gonads. *Journal of Histochemistry & Cytochemistry* 43: 67–76.

Ortega, HH; Palomar, MM; Acosta, JC; Salvetti, NR; Dallard, BE; Lorente, JA; Barbeito, CG & Gimeno, EJ (2008) Insulin-like growth factor I in ovarian follicles and follicular fluid of cows with spontaneous and induced cystic ovarian disease. *Research in Veterinary Science* 84:419–427.

Ortega, HH; Salvetti, NR & Padmanabhan, V (2009) Developmental programming: prenatal androgen excess disrupt ovarian steroid receptor balance. *Reproduction* 137: 865–877.

Paranko, J; Seitz, J & Meinhardt, A (1996) Developmental expression of heat shock protein 60 (HSP60) in the rat testis and ovary. *Differentiation* 60:159-167.

Peter, AT (2004) An update on cystic ovarian degeneration in cattle. *Reproduction in Domestic Animals* 39:1-7.

Powell, E; Wang, Y; Shapiro, DJ & Xu, W (2010) Differential requirements of Hsp90 and DNA for the formation of estrogen receptor homodimers and heterodimers. *Journal of Biological Chemistry* 285:16125-16134.

Pratt, WB & Toft, DO (1997) Steroid receptor interactions with heat shock protein and immunophilin chaperones. *Endocrine Reviews* 18:306-360.

Quintana, FJ & Cohen, IR (2011) The HSP60 immune system network. *Trends in Immunology* 32:89-95.

Ravagnan, L; Gurbuxani, S; Susin, SA; Maisse, C; Daugas, E; Zamzami, N; Mak, T; Jaattela, M; Penninger, JM; Garrido, C & Kroemer, G (2001). Heat-shock protein 70 antagonizes apoptosis-inducing factor. *Nature Cell Biology* 3:839–843.

Rayner, K; Chen, YX; Siebert, T & O'Brien, ER (2010) Heat Shock Protein 27: Clue to understanding estrogen-mediated atheroprotection? *Trends in Cardiovascular Medicine* 20:54-58.

Retzlaff, C; Yamamoto, Y; Hoffman, PS; Friedman, H; Klein, TW (1994) Bacterial heat shock proteins directly induce cytokine mRNA and interleukin-1 secretion in macrophage cultures. *Infection and Immunity* 62:5689–5693.

Romanucci, M; Bastow, T & Della Salda, L (2008) Heat shock proteins in animal neoplasms and human tumours—a comparison. *Cell Stress & Chaperones* 13:253–262.

Ruden, DM; Xiao, L; Garfinkel, MD & Lu, X (2005) Hsp90 and environmental impacts on epigenetic states: a model for the trans-generational effects of diethylstibesterol on uterine development and cancer. *Human Molecular Genetics* 14:R149-155.

Sainis, I; Angelidis, C; Pagoulatos, G & Lazaridis, I (1994) The hsc70 gene which is slightly induced by heat is the main virus inducible member of the hsp70 gene family. *FEBS Letters* 355:282-286.

Salminen, A; Ojala, J; Kaarniranta, K; Hiltunen, M & Soininen, H (2011) Hsp90 regulates tau pathology through co-chaperone complexes in Alzheimer's disease. *Progress in Neurobiology* 93:99-110.

Salvetti, NR; Gimeno, EJ; Lorente, JA & Ortega, HH (2004) Expression of cytoskeletal proteins in the follicular wall of induced ovarian cysts. *Cells Tissues Organs* 178:117–125.

Salvetti, NR; Muller, LA; Acosta, JC; Gimeno, JE & Ortega, HH (2007) Estrogen receptors α and β and progesterone receptors in ovarian follicles of cows with cystic ovarian disease. *Veterinary Pathology* 44:373-378.

Salvetti, NR; Baravalle, C; Mira, GA; Gimeno, EJ; Dallard, BE; Rey, F & Ortega, HH (2009) Heat shock protein 70 and sex steroid receptors in the follicular structures of induced ovarian cysts. *Reproduction in Domestic Animals* 44: 805-814.

Salvetti, NR; Stangaferro, ML; Palomar, MM; Alfaro, NS; Rey, F; Gimeno, EJ & Ortega, HH (2010) Cell proliferation and survival mechanisms underlying the abnormal persistence of follicular cysts in bovines with cystic ovarian disease induced by ACTH. *Animal Reproduction Science* 122:98-110.

Sarkar, S; Arya, R & Lakhotia, SC (2006) Chaperonins: in life and death. In: Sreedhar AS, Srinivas UK (eds) *Stress responses: a molecular biology approach.* Signpost, Trivandrum, India, pp 43–60.

Schmitt, E; Gehrmann, M; Brunet, M; Multhoff, G & Garrido C (2007) Intracellular and extracellular functions of heat shock proteins: repercussions in cancer therapy. *Journal of Leukocyte Biology* 81:15–27.

Silvia, WJ; Hatler, TB; Nugent, AM & Laranja da Fonseca, LF (2002) Ovarian follicular cysts in dairy cows: an abnormality in folliculogenesis. *Domestic Animal Endocrinology* 23:167–177.

Sirotkin, AV & Bauer, M (2011) Heat shock proteins in porcine ovary: synthesis, accumulation and regulation by stress and hormones. *Cell Stress & Chaperones* 16:379–387.

Sreedhar, AS; Kalmar, E; Csermely, P & Shen, YF (2004) Hsp90 isoforms: functions, expression and clinical importance. *FEBS Letters* 562:11-15.

Vanholder, T; Opsomer, G & De Kruif, A (2006) Aetiology and pathogenesis of cystic ovarian follicles in dairy cattle: a review. *Reproduction, Nutrition, Development.* 46:105-119.

Velázquez , MML; Alfaro, NS; Dupuy, CR; Salvetti, NR; Rey, F & Ortega, HH (2010) Heat shock protein patterns in the bovine ovary and relation with cystic ovarian disease. *Animal Reproduction Science* 118 201-209.

Velázquez, MML; Alfaro, NS; Salvetti, NR; Stangaferro, ML; Rey, F; Panzani, C & Ortega, HH (2011) Levels of heat shock proteins transcripts in normal follicles and ovarian follicular cysts. *Reproductive Biology* 11:276-283.

Welch, WJ (1993) How cells respond to stress. *Scientific American* 268:56–64.

Wells, AD; Rai, SK; Salvato, MS; Band, H & Malkovsky, M (1997) Restoration of MHC class I surface expression and endogenous antigen presentation by a molecular chaperone. *Scandinavian Journal of Immunology* 45:605–612.

Woodlock, TJ; Chen, X; Young, DA; Bethlendy, G; Lichtman, MA & Segel, GB (1997) Association of HSP60-like proteins with the L-system amino acid transporter. *Archives of Biochemistry and Biophysics* 338:50–56.

Wrenzycki, C; Lucas-Hahn, A; Herrmann, D; Lemme, E; Korsawe, K; Niemann, H (2002) In vitro production and nuclear transfer affect dosage compensation of the X-linked gene transcripts G6PD, PGK, and Xist in preimplantation bovine embryos. *Biology of Reproduction* 66:127-134.

Wu, BJ; Kingston, RE & Morimoto, RI (1986) Human HSP70 promoter contains at least two distinct regulatory domains. *Proceedings of the National Academy of Sciences of the United States of America* 83:629–633.

Yamagishi, N; Saito, Y; Ishihara, K & Hatayama, T (2002) Enhancement of oxidative stress-induced apoptosis by Hsp105alpha in mouse embryonal F9 cells. *European Journal of Biochemistry* 269:4143-4151.

Yuan, J; Xiao, L; Lu, C; Zhang, X; Liu, T; Chen, M; Hu, Z; Gao, F & Liu, Y (2009) Increased expression of heat shock protein 105 in rat uterus of early pregnancy and its significance in embryo implantation. *Reproductive Biology and Endocrinology* 7:23.

Zhang, XS; Lue, YH; Guo, SH; Yuan, JX; Hu, ZY; Han, CS; Hikim, AP; Swerdloff, RS; Wang, C & Liu, YX (2005) Expression of HSP105 and HSP60 during germ cell apoptosis in the heat-treated testes of adult cynomolgus monkeys (macaca fascicularis). *Frontiers in Bioscience* 10:3110-3121.

In: Heat Shock Proteins
Editor: Saad Usmani
ISBN: 978-1-62417-571-8

Chapter IV

The Role of Heat Shock Proteins in Huntington's Disease

Leigh Anne Swayne and Joana Gil-Mohapel*
Division of Medical Sciences, Island Medical Program, University of Victoria, Victoria, British Columbia, Canada

Abstract

Huntington's disease (HD) is the most common polyglutamine neurodegenerative disorder. The mutation consists of an unstable expansion of CAG repeats within the coding region of the *HD* gene, which expresses the protein huntingtin. Although the abnormal protein is ubiquitously expressed throughout the organism, cell degeneration occurs mainly in the brain, particularly in the striatum and cortex. The first symptoms usually appear in mid-life and include cognitive deficits and motor disturbances associated with the loss of voluntary movement coordination, which progress over time. The disease is invariably fatal and there is currently no cure for affected individuals. At the neuropathological level, HD is characterized by the formation of ubiquitinated neuronal intranuclear inclusions (NIIs) of mutant huntingtin. NIIs recruit and seed the aggregation of several intracellular proteins including molecular chaperones such as the heat shock proteins HSP40 and HSP70. Since the abnormally long polyglutamine tract present in the mutant protein is likely to result in an overall misfolding of huntingtin, the interaction of HSPs with mutant huntingtin may represent an initial attempt of the cell to refold the mutant protein. However, by being sequestered into aggregates, these chaperones are prevented from exerting their normal protective functions and, over the course of the disease, this is likely to result in the overwhelming accumulation of misfolded proteins inside susceptible neurons. In this book chapter we present an overview of recent evidence implicating the failure of endogenous molecular chaperones to protect against the neuropathological mechanisms underlying HD. Furthermore, as inhibition of misfolding and oligomerization of mutant huntingtin is of potential

* Corresponding Author: Joana Gil-Mohapel, Ph.D., Division of Medical Sciences, Island Medical Program, University of Victoria, Victoria, B.C., V8W 2Y2, Canada. E-mail: jgil@uvic.ca; Phone: +1.250-721-6586; Fax: +1.250-472-5505.

therapeutic value, we will also discuss the impact of modulating the expression of these molecular chaperones, in particular HSPs, for the treatment of HD.

Keywords: Aggregates; chaperones; heat shock proteins; Huntington's disease; misfolding; neuronal intranuclear inclusions; oligomerization; synaptic function

List of Abbreviations

CAG, cytosine-adenine-guanine; *C. elegans, Caenorhabditis elegans;* CHIP, C-terminal HSP70 interacting protein; CREB, cyclic-AMP response element-binding protein; CSPα, cysteine string protein alpha; DARPP-32, dopamine- and cyclic-AMP-regulated phosphoprotein of 32 kDa; EGFP, enhanced green fluorescent protein; GABA, gamma-aminobutyric acid; GAK, Cyclin G-associated kinase; GFP, green flourescent protein; HD, Huntington's disease; HEAT, **H**untingtin, **E**longation factor3, **A** subunit of protein phosphatase 2A, **T**OR1; HSC, heat shock cognate chaperone; HSF1, heat shock transcription factor 1; HSP, heat shock proteins; KO, knock-out; NIIs, neuronal intranuclear inclusions; NMDA, *N*-methyl-D-aspartate; SCA, spinocerebellar ataxia; SGT, tetratricopeptide repeat (TPR)-containing protein; SNARE, **S**oluble **N**-ethylmaleimide sensitive factor **A**ttachment Protein **RE**ceptor; Sp1, specific protein-1; synprint, **syn**aptic **pr**otein **int**eraction; $TAF_{II}130$, TATA-binding protein associated factor; YAC, yeast artificial chromosone; YFP, yellow fluorescent protein; 17-AAG, 17-(allylamino)-17-demethoxygeldanamycin; 17-DMAG, 17-(dimethylaminoethylamino)-17-demethoxygeldanamycin.

1. Huntington's Disease

Huntington's disease (HD) is the most common polyglutamine neurodegenerative disorder with a prevalence of 3 to 10 affected subjects per 100,000 individuals in Western Europe and North America (for review see (Vonsattel and DiFiglia, 1998). The disease is caused by an unstable expansion of cytosine-adenine-guanine (CAG) repeats within the coding region of the *HD* gene, which encodes the protein huntingtin. This mutation results in a stretch of glutamine residues located in the NH_2-terminus of huntingtin (The Huntington's Disease Collaborative Research, 1993).

In most cases the onset of the disease occurs in midlife, between the ages of 35 and 50 years. The disease progresses over time and is invariably fatal 15 to 20 years after the onset of the first symptoms. Motor disturbances, associated with the loss of voluntary movement coordination, are the classical symptoms of HD, with bradykinesia and rigidity appearing in later stages of the disease. Cognitive capacities are also severely affected during the course of the disease with the slowing of intellectual processes being the first sign of cognitive impairment in HD patients (for review see Gil and Rego, 2008). In fact, deficits in some cognitive functions can in some cases be detected decades before the onset of motor symptoms. These cognitive impairments worsen over time and late-stage HD patients show profound dementia (Folstein et al., 1983). Moreover, psychiatric symptoms including personality changes, irritability, apathy, depression, and anxiety are also common (Dewhurst et al., 1970; Duff et al., 2007; Folstein et al., 1983; Jensen et al., 1993; Vaccarino et al., 2011; van Duijn et al., 2007; van Duijn et al., 2010).

Mutant huntingtin is ubiquitously expressed throughout the body. However, cell degeneration occurs mainly in the brain, particularly in the striatum and certain layers of the cortex (for review see Gil and Rego, 2008; Vonsattel and DiFiglia, 1998). Nevertheless, cell loss can also be detected in other brain regions, including the hippocampus and the lateral hypothalamus (Petersen et al., 2005; Rosas et al., 2003; Spargo et al., 1993).

Various mechanisms have now been implicated in the pathogenesis of HD (for review see Gil and Rego, 2008). At the neuronal level, both a loss of function of normal huntingtin and a toxic gain of function conferred by the polyglutamine expansion are thought to contribute to HD pathology, promoting the activation of proteases such as caspases and calpains, transcription deregulation, metabolic and mitochondrial dysfunction, oxidative stress, synaptic dysfunction, inhibition of protein degradation, and protein misfolding and oligomerization, which over the years culminate in neurodegeneration and cell loss in susceptible brain regions (for review see Gil and Rego, 2008).

2. Synaptic Function in Huntington's Disease

Huntingtin is a very large protein (350 kDa) that contains numerous HEAT repeats (**H**untingtin, **E**longation factor 3, **A** subunit of protein phosphatase 2A and **T**OR1) (Andrade and Bork, 1995). The HEAT repeat, a motif of 40 amino acids involved in mediating protein-protein interactions, is found in a variety of proteins involved in intracellular transport (Neuwald and Hirano, 2000; Takano and Gusella, 2002). Indeed, huntingtin participates in a large number of protein-protein interactions (Harjes and Wanker, 2003; Kaltenbach et al., 2007; Li and Li, 2004). Its interaction partners are involved in cellular functions ranging from intracellular transport to regulation of calcium (Ca^{2+}) homeostasis. These diverse protein-protein interactions appear to underlie the function of wild-type huntingtin.

Huntingtin is also modulated by post-translational processing events, in particular proteolytic cleavage. Both its protein interactions and proteolytic processing can be altered by polyglutamine expansion, which makes the regulation of huntingtin protein levels an important aspect of HD etiology and pathophysiology. Importantly, polyglutamine expansion appears to alter the functions of several huntingtin interaction parters and this is thought to play a key role in HD pathogenesis (Gafni et al., 2004; Hermel et al., 2004; Li et al., 2000; Wellington et al., 2002). Furthermore, the proteolytic cleavage of huntingtin by calpains (Gafni and Ellerby, 2002; Gafni et al., 2004; Kim et al., 2001) and caspases (Goldberg et al., 1996; Hermel et al., 2004; Martindale et al., 1998; Wellington et al., 2002; Wellington et al., 1998; Wellington et al., 2000), and the consequent generation of huntingtin N-terminal fragments of various sizes (~60-80 kDa), is enhanced in HD brains. These mutant huntingtin N-terminal fragments have been shown to associate with synaptic vesicles (Li et al., 2000). In line with this synatic localization, many of the neurotoxic mechanisms elicited by mutant huntingtin involve synaptic dysfunction (Rozas et al., 2010; Smith et al., 2005)(Cepeda et al., 2003; Cummings et al., 2009; Rozas et al., 2011; Smith et al., 2005; Zeron et al., 2002).

How does mutant huntingtin affect synapses? Mutant huntingtin impacts on several aspects of Ca^{2+} homeostasis and directly impairs the cellular machinery involved in synaptic transmission both at the pre- and post-synapses. In addition to the important effects of mutant huntingtin on post-synaptic *N*-methyl-D-aspartate (NMDA) receptor function (reviewed in Fernandes and Raymond, 2009), mutant huntingtin also dysregulates pre-synaptic neurotransmitter release (Romero et al., 2008; Rozas et al., 2010, 2011; Swayne et al., 2005). Several recent studies link mutant huntingtin to changes in the expression and regulation of members of the synaptic vesicle fusion SNARE (**S**oluble **N**-ethylmaleimide sensitive factor **A**ttachment Protein **RE**ceptor) protein complex. For example, huntingtin N-terminal fragments disengage important coupling between pre-synaptic N-type Ca^{2+} channels and SNAREs (Swayne et al., 2005).

In addition to their critical role in synaptic vesicle fusion and neurotransmitter release, SNAREs also regulate Ca^{2+} fluxes that are responsible for triggering neurotransmitter release. The binding of one of the SNAREs, syntaxin, to the **syn**aptic **pr**otein **int**eraction (synprint) site of N-type Ca^{2+} channels causes a hyperpolarizing shift in steady-state inactivation, which increases G-protein inhibition. These alterations result in reduced Ca^{2+} channel availability and Ca^{2+} entry (Bezprozvanny et al., 1995; Jarvis et al., 2002; Jarvis et al., 2000; Jarvis and Zamponi, 2001; Leveque et al., 1994; Sheng et al., 1996; Stanley and Mirotznik, 1997; Wiser et al., 1996). Swayne et al. demonstrated that huntingtin N-terminal fragments interact with the synprint site, thereby competing with syntaxin for synprint binding and robustly reducing syntaxin regulation of these channels. Thus, huntingtin N-terminal fragments increase Ca^{2+} influx through N-type Ca^{2+} channels (Swayne et al., 2005). As a result, the increased generation of huntingtin N-terminal fragments caused by polyglutamine expansion has the potential of playing a crucial role in synaptic perturbations in HD by modulating cross-talk between these two essential proteins. A recent elegant paper by Rozas et al. (Rozas et al., 2011) provided *in vivo* support for this hypothesis, demonstrating that evoked neurotransmitter release is increased in the R6/1 HD mouse model [which expresses exon 1 of the human HD gene with approximately 115 CAG repeats (Mangiarini et al., 1996)] as compared to controls. Whether this phenomenon is common to all HD models and human HD remains to be determined.

3. The Role of Heat Shock Proteins in Synaptic Function

Several types of HSPs play specific, critical roles in synaptic function beyond their generalized stereotypical roles in cell survival and maintenance (for a recent review see Stetler et al., 2010). As their name implies, the HSPs were orignally detected in the cellular response to heat and other stress factors (Ashburner and Bonner, 1979; Craig, 1985; Spradling et al., 1975). The existence of highly homologous constitutively expressed relatives, heat-shock cognate proteins (HSCs), which perform critical cellular housekeeping functions, is now also generally recognized. It should be noted that due to the sequence and functional similarities, the term HSP now refers to both types of molecular chaperones.

Rapid Ca^{2+}-evoked pre-synaptic neurotransmitter release needs to be both precise and durable. Furthermore, given the spatial separation between the synapse and the cell body, the synapse must be able to operate with a certain level of independence. Thus, it is not surprising that a special throng of hard-working proteins, the molecular chaperones, are necessary to sustain synaptic transmission.

Molecular chaperones are classically defined as proteins that recognize folding intermediates of proteins and stabilize them during polypeptide folding, assembly, and disassembly (for reviews see Gething and Sambrook, 1992; Young et al., 2004). The major chaperone system controlling synapse stability is the HSP70/DNAJ co-chaperone system (Cheetham and Caplan, 1998; Cyr et al., 1994; Johnson et al., 2010; Qiu et al., 2006; Zinsmaier, 2010; Zinsmaier and Bronk, 2001). These important synaptic players protect synapses from the putative deleterious effects of aberrantly modified/folded and/or functionally impaired proteins (such as mutant huntingtin).

HSP proteins transiently bind and release unfolded regions of substrate proteins using energy garnered by ATP hydrolysis (Pelham, 1986; Young et al., 2003). The ATPase and protein binding capacities of the HSP's can be modulated by protein-protein interactions with other proteins. These are commonly referred to as co-chaperones. A co-chaperone family of major importance for synaptic function and stability are the DNAJ proteins (Braun and Scheller, 1995; Braun et al., 1996; Johnson et al., 2010). DNAJ proteins are so named because they contain J-domains. DNAJ proteins stimulate the HSP ATPase activity (Braun et al., 1996; Chamberlain and Burgoyne, 1997) and are often targeted to specific intracellular locations, to which they subsequently recruit cytosolic HSPs. This allows for the conformational modulation of specific target proteins at appropriate intracellular sites (for reviews see Echtenkamp and Freeman, 2011; Johnson, 2011; Qiu et al., 2006; Young et al., 2003). Thus far, eight J-domain-containing proteins have been identified in neurons. Three of these are vital at the synapse: auxilin, cyclin G-associated kinase (GAK), and the cysteine string protein alpha (CSPα). Auxilin recruits HSC70 to uncoat clathrin-coated vesicles together with GAK. CSPα recruits HSC70 and small glutamine-rich tetratricopeptide repeat (TPR)-containing protein (SGT) to maintain Ca^{2+}-triggered neurotransmitter release (Braun and Scheller, 1995; Braun et al., 1996; Chamberlain and Burgoyne, 1997; Tobaben et al., 2003).

3.1. Modulation of the Pre-Synapse by Chaperones and Huntingtin

Together, CSPα (Braun et al., 1996; Chamberlain and Burgoyne, 1997) and SGT (Tobaben et al., 2001; Tobaben et al., 2003) maximally stimulate the chaperone (ATPase) activity of HSC70. CSPα was originally discovered in *Drosophila melanogaster* as a synaptic antigen (Zinsmaier et al., 1990) and soon after in *Torpedo* as a subunit of pre-synaptic N-type Ca^{2+} channels (Gundersen and Umbach, 1992). CSPα is expressed at most synapses of the central nervous system and at neuro-muscular junctions on synaptic vesicles (~1% of total synaptic vesicle protein). It is also expressed on secretory vesicles of endocrine, neuroendrocine, and exocrine cells. Following its discovery, the ensuing investigation into the specific role(s) of CSPα in secretion initially generated heated debate, in part due to the different speed requirements of the various secretory systems, as well as the different types of Ca^{2+} channels involved in evoked release.

This early work together with recent CSPα knock-out (KO) mouse studies support a hypothesis that the CSPα chaperone system controls the pre-synaptic release machinery

SNARE complex as well as the assembly and modulation of pre-synaptic Ca^{2+} channels. Given the synaptic vesicle localization of CSPα, the CSPα/HSC70/SGT chaperone complex was originally proposed to carry out a chaperone function in neurotransmitter release (Bronk et al., 2001; Tobaben et al., 2001; Tobaben et al., 2003). It was postulated that a chaperone system would be required on the premise that proteins become misfolded by their heavy use in rapid and continuous cycles of exocytosis and endocytosis. In the current working model (Zinsmaier, 2010), CSPα recruits HSC70 and SGT to synaptic vesicles, which is followed by ATP hydrolysis-driven renaturation/reactivation of synaptic substrate proteins. In agreement with this model, deletion of CSPα results in considerable neurodegeneration leading to decreased viability of mice (Fernandez-Chacon et al., 2004), and flies (Zinsmaier et al., 1994). CSPα mutant *Drosophila* first elucidated the critical role of this protein in synaptic stability, including thermal protection, Ca^{2+}-triggered synaptic vesicle fusion and/or pre-synaptic Ca^{2+} homeostasis (both at rest and during high frequency stimulation), as well as synaptic growth. Accordingly, ensuing studies in CSPα KO mice revealed progressive motor and sensory deficits, paralysis, blindess, and premature death during early adulthood (between postnatal days 40-60) (Chandra et al., 2005; Fernandez-Chacon et al., 2004; Schmitz et al., 2006). Interestingly, evoked neurotransmitter release is normal in young CSPα KO mice but later becomes asynchronous and progressively worsens with age. Interestingly, these impairments were resolved by increasing extracellular Ca^{2+} (Ruiz et al., 2008; Dawson-Scully et al., 2000), pointing to a reduced ability of Ca^{2+}-triggered exocytosis in the absence of CSPα.

CSPα interacts with individual SNAREs and SNARE complexes (Nie et al., 1999; Swayne et al., 2006; Wu et al., 1999). Interestingly, the levels of pre-synaptic SNARE complexes and the SNARE protein **syn**aptosomal-**a**ssociated **p**rotein **25** (SNAP-25) are lower in the absence of CSPα (Sharma et al., 2011). Furthermore, the interaction of CSPα with SNARE complexes does not preclude concomitant binding to HSC70 (Swayne et al., 2006) and pre-synaptic N-type Ca^{2+} channels (Magga et al., 2000; Miller et al., 2003b; Swayne et al., 2006). The binding of CSPα to N-type Ca^{2+} channels and Gβγ produces a tonic G-protein inhibition of these channels (Magga et al., 2000; Miller et al., 2003b). Interestingly, this important N-type Ca^{2+} channel regulation is disrupted by mutant huntingtin (Miller et al., 2003a).

A recent elegant study demonstrated that binding of the CSPα/HSC70/SGT complex to monomeric SNAP-25 is required to prevent its aggregation, which thus allows subsequent assembly of SNARE complexes (Sharma et al., 2011). Aberrant folding of SNAP-25 in CSPα KO animals prevents its incorporation into SNARE complexes and targets it for proteasomal degradation. This control over SNAP-25 stability is dynamically regulated by the levels of CSPα. Thus, CSPα-dependent chaperone activity plays a vital role in SNAP-25 and the function of the SNARE complex during rapid cycles of exocytosis and endocytosis. Interestingly, the levels of SNAP-25 are reduced in human HD brains (Smith et al., 2007) and mutant huntingtin aggregates have been shown to sequester SNARE proteins (Swayne and Braun, 2007), which together might contribute to the synaptic dysfunction observed in HD.

The ability of mutant huntingtin to cause synaptic dysfunciton by disrupting CSPα/HSC70/SGT synaptic chaperone activity is in agreement with accumulating evidence for the overpowering of synaptic molecular chaperone systems by mutant huntingtin (Hay et al., 2004; Tagawa et al., 2007; Yamanaka et al., 2008; Zinsmaier, 2010). Furthermore, new data suggest that neuron subtype-specific differences in HSP induction might underlie the selective vulnerability of specific neuronal populations in HD (Tagawa et al., 2007).

On the other hand, the identification of compounds that can selectively modulate the neuroprotective functions of CSPα may be of therapeutic value for the treatment of neurodegenerative diseases such as HD, by preventing synaptic dysfunction. In line with this hypothesis, a recent study by Xu and collaborators (2010) has shown that quercetin, a plant-derived flavonoid thought to prevent memory loss and altitude sickness, can selectively modulate CSPα's neuroprotective function by regulating the formation of stable CSPα-CSPα dimers (Xu et al., 2010). Similarly, other compounds that selectively target J-proteins may also have considerable potential as novel therapeutic agents for neurodegenerative diseases such as HD that are characterized by synaptic dysfunction and future studies are thus warranted in order to test this hypothesis.

4. Protein Aggregation in Huntington's Disease

One of the neuropathological hallmarks of HD is the intracellular accumulation of mutant huntingtin with the formation of insoluble ubiquitinated neuronal intranuclear inclusions (NIIs) and protein aggregates in dystrophic neurites (DiFiglia et al., 1997). Evidence from both *in vitro* (Arrasate et al., 2004; Saudou et al., 1998) and *in vivo* (Slow et al., 2005) models has suggested that the formation of NIIs may have a neuroprotective function by sequestering toxic N-terminal fragments and oligomers of mutant huntingtin as well as other misfolded proteins, which in their soluble forms could cause a more rapid and severe damage (for review see Ciechanover and Brundin, 2003; Gunawardena and Goldstein, 2005; Gil and Rego, 2008).

However, it is likely that with the progression of the disease, an increase in aggregation might contribute to neuronal dysfunction in the affected regions of the HD brain (for review see Gil and Rego, 2008). Indeed, several transcription factors including the cyclic-AMP response element-binding protein (CREB)-binding protein (Steffan et al., 2000; Schaffar et al., 2004), specific protein-1 (Sp1) (Li et al., 2002; Dunah et al., 2002), and the TATA-binding protein associated factor ($TAF_{II}130$) (Dunah et al., 2002) have all been shown to co-localize with NIIs, thus contributing to transcriptional dysregulation in HD. Furthermore, an increase in aggregation can also cause the dysregulation of axonal transport by physically blocking narrow axonal terminals (for review see Gunawardena and Goldstein, 2005). Finally, inclusions may physically block the proteasome, preventing the entrance of further substrates into this protein degradation complex (for review see Goellner and Rechsteiner, 2003), while also sequestering molecular chaperones such as HSPs (Wyttenbach et al., 2000; Qin et al., 2004). This might result in an overall increase in the intracellular content of misfolded proteins that are not being refolded or degraded. As such, many authors hypothesize that aggregation of mutant huntingtin can result in a severe deregulation of several key intracellular pathways that eventually culminate in neuronal demise in the HD brain (for review see Gil and Rego, 2008).

5. The Role of Heat Shock Proteins in Protein Aggregation in Huntington's Disease

HSPs are expected to play a protective role in HD, as they assist the folding of proteins into appropriate conformations, refold misfolded proteins, and rescue previously aggregated proteins (for review see Glover and Lindquist, 1998; Hartl, 1996; Hendricks and Hartl, 1993). In order to test this hypothesis, various studies have now investigated how the expression of certain chaperones can alter huntingtin aggregation using *in vitro* models of HD.

In an initial study, Wyttenbachand collaborators (2000) investigated the effects of over-expressing the human HSP40 homologue HDJ-2/HSDJ, a co-chaperone that regulates the activity of members of the DnaK (HSP70) family (for review see Hartl, 1996; Hendricks and Hartl, 1993), on polyglutamine-induced aggregation *in vitro* (Wyttenbach et al., 2000). By transiently transfecting constructs of huntingtin exon 1 with different repeat lengths in COS-7, PC12, and SH-SY5Y cells, the authors confirmed that aggregation was repeat length-dependent and that HSP70, HSP40, the 20S subunit of the proteasome, and ubiquitin co-localized with inclusions. Furthermore, induction of a heat-shock response and treatment with the proteasome inhibitor lactacystin increased the proportion of cells with inclusions. Surprisingly however, over-expression of HDJ-2/HSDJ did not reduce inclusion formation in PC12 and SH-SY5Y cells and actually resulted in an increase in inclusion load in COS-7 cells (Wyttenbach et al., 2000). The apparent contradiction between these results and the proposed role of HSPs in protein refolding may reflect a dual role for these chaperones in polyglutamine aggregation and the net effect of HSPs on this process may depend on the overall balance between different types of HSPs.

In a subsequent study, Sittler and colleagues (2001) have tested the effects of geldanamycin, a benzoquinone ansamycin that binds to HSP90 and activates a heat shock response in mammalian cells in a cell culture model of HD [COS-1 cells expressing a construct of enhanced green fluorescent protein (EGFP) fused with huntingtin exon 1 carrying 72 CAG repeats (HD72Q)]. Geldanamycin induced the expression of HSP40, HSP70 and HSP90 and inhibited aggregation of huntingtin fragments in a similar fashion as over-expression of HSP70 and HSP40, showing that huntingtin aggregation can be supressed by inducing the activation of a heat shock response *in vitro* (Sittler et al., 2001). Despite these promising results, geldanamycin possesses toxic and unfavorable pharmacokinetic properties and therefore it is not suitable for clinical use (Samuni et al., 2010; Wu et al., 2010). However, its pharmacologically improved derivative 17-(allylamino)-17-demethoxygel-danamycin (17-AAG) and its water-soluble analogue 17-(dimethylaminoethylamino)-17-demethoxygeldanamycin (17-DMAG) can also induce the expression of the molecular chaperones HSP40, HSP70, and HSP105 in mammalian cells and inhibit the formation of mutant huntingtin aggregates with higher efficiency than geldanamycin itself. Therefore, these geldanamycin derivatives constitute potential pharmacological candidates for the treatment of polyglutamine disorders involving protein aggregation such as HD (Herbst and Wanker, 2007). Similar results were also obtained with a purified preparation of HSP70 /HSC70 from bovine muscle (Novoselova et al., 2005). In this study, human neuroblastoma SK-N-SH cells were transfected with huntingtin exon 1 carrying 103 CAG trinucleotide repeats coupled to green flourescent protein (GFP), leading to the formation of insoluble protein aggregates and cell death through apoptosis. Treatment with HSP70-HSC70 resulted in a 40-50% reduction in the number of apoptotic cells, as well as a significant decrease in the number and size of huntingtin inclusions (Novoselova et al., 2005).

Other chaperones have also been shown to inhibit huntingtin aggregation *in vitro*. One example is the C-terminal HSP70 interacting protein (CHIP), both a co-chaperone and ubiquitin ligase that plays an important role in protein quality control by promoting protein refolding through molecular chaperones and protein degradation through the ubiquitin-proteasome system (Connell et al., 2001; Meacham et al., 2001; Kampinga et al., 2003). Indeed, CHIP was shown to suppress polyglutamine aggregation and toxicity in transfected cell lines (COS-7 and PC-12 cells), primary neurons, and a zebrafish model of HD. Interestingly, aggregation inhibition by CHIP is dependent upon its co-chaperone function, suggesting that CHIP acts to facilitate the solubility of enlongated polyglutamine residues through its interactions with molecular chaperones. Furthermore, N171-82Q HD transgenic mice [which express a cDNA encoding an N-terminal fragment of huntingtin with 171 amino acids bearing 82 glutamine residues; (Schilling et al., 1999)] that are haploinsufficient for CHIP display a markedly accelerated disease phenotype, with an earlier onset of motor symptoms, a decrease in life span and a significant increase in the number of huntingtin aggregates and inclusions, clearly indicating that CHIP is a critical mediator of the neuronal response to misfolded enlongated polyglutamine residues as in the case of HD (Miller et al., 2005c).

Together, the results from these *in vitro* studies provide evidence that molecular chaperones in general and HSPs in particular are modulators of protein aggregation and toxicity in HD.

6. Over-Expression of Heat Shock Proteins as a Therapeutic Strategy for Huntington's Disease

Since HSPs have proven to decrease protein aggregation *in vitro* (Section 5), over-expression of these proteins and/or induction of a heat shock response might be of therapeutic value to HD by rescuing aggregation. As such, several studies have tested whether similar results could be obtained in various HD *in vivo* models.

6.1. Non-Mammalian Models

I*n vivo* non-mammalian HD models, such as those generated in *Caenorhabditis elegans (C. elegans)* and *Drosophila*, have proven to be valuable in studying important intracellular pathways involved in the disease process (including aggregation mechanisms). These models are also useful for rapidly screening potentially beneficial pharmacologic compounds (such as inhibitors of aggregation) (for review see Marsh et al., 2003).

C. elegans Models. Hydroxylamine derivatives act as co-inducers of HSPs and can enhance the expression of these molecular chaperones in diseased cells, without significant adverse effects (Chung et al., 2008; Kalmar et al., 2008; Kieran et al., 2004; Vígh et al., 2007). As such, the potential beneficial effects of NG-094, a novel hydroxylamine derivative, have recently been tested in a *C. elegans* model that expresses polyglutamine expansions with

35 residues fused to the yellow fluorescent protein (Q35-YFP) in the muscle cells of the body wall (Haldimann et al., 2011). NG-094 administration significantly ameliorated polyglutamine-mediated animal paralysis, reduced the number of Q35-YFP aggregates, and delayed the polyglutamine-dependent acceleration of aging, even when the drug was administered after disease onset. Importantly, NG-094 was shown to confer cytoprotection through a mechanism involving the heat-shock transcription factor HSF-1, which resulted in an overall increase in the expression of stress-inducible HSPs (Haldimann et al., 2011). Therefore, the results from this study indicate that NG-094 is a promising therapeutic candidate for neuropathologies that involve the formation of polyglutamine-containing aggregates as in the case of HD and future preclinical studies on mammalian models of polyglutamine disease are thus warranted.

Drosophila melanogaster Models. Two studies have also tested the effects of modulating the expression of HSPs in *Drosophila* models of HD. In the first study a well establised *Drosophila* model that expresses exon 1 of the human *HD* gene with 93 CAG repeats (Steffan et al., 2001) was used to test and compare the therapeutic efficacies of either HSP70 over-expression or intrabody therapy in ameliorating the *Drosophila* HD phenotype (McLear et al., 2008). Intrabodies are intracellular single-chain Fv and single-domain antibodies that have the potential to alter the folding, interactions, modifications, and/or the subcellular localization of their targets (Miller et al., 2005a; Messer et al., 2006). Importantly, intrabody therapy has previously been shown to reduce aggregation (Lecerf et al., 2001; Colby et al., 2004; Miller et al., 2005b) and cell toxicity in *in vitro* models of HD (Colby et al., 2004; Miller et al., 2005b; Murphy et al., 2004) as well as to suppress neuropathology in a *Drosophila* HD model (Wolfgang et al., 2005) and in yeast artificial chromosone (YAC) 128 HD transgenic mice (Southwell et al., 2009), which express the full-length human huntingtin gene with approximately 128 CAG repears (Slow et al., 2003). In this study, over-expression of human HSP70 in this *Drosophila* HD model resulted in improved survival and prolonged adult life when compared with intrabody treatment alone and an additive effect on adult survival was observed when the two therapies were combined. However, intrabody therapy was more successful at suppressing photoreceptor neurodegeneration and blocking aggregation of mutant huntingtin than HSP70. Nevertheless, suppression of endogenous HSP70 expression resulted in increased pathology, clearly indicating that this HSP plays an important role in suppressing mutant huntingtin-induced toxicity (McLear et al., 2008). Together, the results from this study suggest that different neuronal populations may be more responsive to either HSPs or intrabodies, and therefore, a combination of both is likely to show enhanced therapeutic benefits and preclinical trials using HD transgenic mouse models are thus warranted in order to test this exciting hypothesis.

The second study assessed the therapeutic effects of the geldanamycin derivative 17-AAG, which was previously shown to induce the activation of multiple molecular chaperones (Herbst and Wanker, 2007; Section 4.1), on polyglutamine-induced neurodegeneration in *Drosophila* models of various polyglutamine diseases including spinocerebellar ataxia (SCA) and HD (Fujikake et al., 2008). This compound was found to suppress neurodegeneration in both the SCA and the HD models through a mechanism that is depedent upon the expression of HSP70, HSP40, and HSP90. Importantly, knock-down of the heat shock transcription factor 1 (HSF1) abolished the induction of HSPs and the therapeutic effect of 17-AAG, indicating that the beneficial effects of this compound are mediated by HSF-1-mediated activation of HSPs. These results confirm the intial *in vitro* studies (Herbst and Wanker,

2007; Section 4.1) and indicate that 17-AAG is a promising therapeutic candidate for HD (Fujikake et al., 2008).

6.2. Rodent Models

Modeling HD in rodent models is crucial for the understanding of the relationships between neuronal dysfunction and/or death and the development of abnormal behaviours. Moreover, rodent models also provide excellent tools to further test the efficacy of new therapies and compounds, including those originally screened with the simpler *C. elegans* and *Drosophila* models (for review see Beal and Ferrante, 2004; Hersch and Ferrante, 2004; Li JY et al., 2005; Gil and Rego, 2009; Gil-Mohapel et al., 2011).

Lentiviral-Based Rat Model. A recent study has used an HD rat model based on lentiviral-mediated over-expression of the first 171 amino acids of the human huntingtin protein with 82 glutamine repeats to test the effects of over-expression of the molecular chaperones HSP104 and HSP27 on huntingtin-induced striatal neurogeneration (Perrin et al., 2007). Over-expression of both chaperones was shown to significantly reduce mutant huntingtin-induced striatal loss of dopamine- and cyclic-AMP-regulated phosphoprotein of 32 kDa (DARPP-32) expression (a marker of striatal neurons). Furthermore, HSPs also altered the distribution and size of NIIs, although no decrease in the actual number of NIIs was detected. Interestingly, the expression of the mutant huntingtin fragment was enough to increase the expression of endogenous HSP70, which appeared co-localized with NIIs. On the other hand, the over-expression of HSP104 and HSP27 modified the sub-cellular localization of HSP70 from the NIIs to the cytoplasm while also inducing the expression of endogenous HSP27, resulting in an overall increase in the amount of cytoplasmic HSPs (Perrin et al., 2007).

Transgenic Mouse Models. To determine whether overexpression of HSPs could have beneficial effects in a transgenic mouse model of HD, Hansson and co-workers crossed R6/2 mice [which express exon 1 of the human *HD* gene with approximately 150 CAG repeats (Mangiarini et al., 1996)] with mice over-expressing HSP70 (Hansson et al., 2003). However, over-expression of this molecular chaperone had no effect on survival, development of paw clasping, and neuropathology (as assessed by brain weight, striatal atrophy, striatal neuronal atrophy, and number and size of NIIs) of R6/2 mice. Although highly expressed in the hippocampus, cortex and striatum, the cytoplasmic levels of HSP70 seemed reduced in many neurons, with a clear sequestration into NIIs (Hansson et al., 2003). In agreement with this study, cross-breeding R6/2 mice with mice overexpressing HSP27, had also no effect on the development of the R6/2 HD phenotype (Zourlidou et al., 2007).

More recently, pharmacological activation of HSF-1 was shown to improve huntingtin aggregate load, motor performance, and other HD-related phenotypes in R6/2 mice (Labbadia et al., 2011). However, these beneficial effects were only transient and diminished with disease progression. Interestingly, this transient effect seems to be related with altered chromatin architecture, reduced HSF-1 binding, and an overall impaired heat shock response that are aggravated by disease progression in the R6/2 HD transgenic mouse model. Importantly, similar findings were also obtained with a different HD mouse model, the HdhQ150 knock-in mice (Labbadia et al., 2011), and indicate that the mechanisms underlying the induction of a normal heat shoch response are disrupted in genetic models of HD and that the pharmacological induction of HSF-1 as a therapeutic approach to HD is more complex than was previously anticipated.

As such, despite their well-established role as modulators of protein misfolding and consequent aggregation and the promising results obtained in cell cutures (Section 5), *C. elegans*, *Drosophila* (Section 6.1), and lentiviral-based rat (Section 6.2) models of HD, over-expression of HSPs may delay polyglutamine toxicity only up to a certain level in genetic models of HD. However, these results do not rule out the possibility that targeting the heat shock response system in combination with other pharmacological interventions (that can prevent the disease-induced impairment of the normal heat shock response and/or have beneficial effects at the level of the proteosome system or transcriptional dysregulation) might still have an impact on aggregation in these genetic models of HD (Jackrel and Shorter, 2011).

Conclusion

The heat shock response is a highly conserved protective mechanism that enables cells to withstand diverse environmental stressors that disrupt protein homeostasis while also promoting protein misfolding. This system has evolved to become a key regulator of synaptic speed and integrity. As follows, pertubances to synaptic chaperone functioning are linked to several neurodegenerative disorders, including HD. Moreover, it has been suggested that pharmacological strategies that can elicit a heat shock response (for example by activating the transcription factor HSF-1) might help mitigate protein misfolding and aggregation in neurodegenerative disorders where aberrant protein aggregation occurs. In agreement, a variety of complementary approaches that used immortalized cell lines, primary neuronal cultures, as well as *C. elegans* and *Drosophila* models of HD have shown that increasing the expression of several HSPs could rescue aggregation of mutant huntingtin as well as neuronal degeneration. However, such beneficial effects were not replicated in genetic mouse models, which more closely mimick the human condition. As such, a synergistic combination of therapies is likely to be needed in order to fully restore protein homeostasis in HD.

Acknowledgments

LAS acknowledges funding from the University of Victoria Start-Up Fund, the Victoria Foundation, as well as the Natural Sciences and Engeneering Research Council of Canada (NSERC) Discovery Grant Program. JGM acknowledges the Michael Smith Foundation for Health Research (MSFH; Canada) for post-doctoral funding.

References

Andrade, M. A. & Bork, P. (1995). HEAT repeats in the Huntington's disease protein. *Nat Genet*, 11, 115-116.

Arrasate, M., Mitra, S., Schweitzer, E. S., Segal, M. R. & Finkbeiner, S. (2004). Inclusion body formation reduces levels of mutant huntingtin and the risk of neuronal death. *Nature*, 431, 805-810.

Ashburner, M. & Bonner, J.J. (1979). The induction of gene activity in drosophilia by heat shock. *Cell*, 17, 241-254.

Beal, M. F. & Ferrante, R. J. (2004). Experimental therapeutics in transgenic mouse models of Huntington's disease. *Nat Rev Neurosci*, 5, 373-384.

Bezprozvanny, I., Scheller, R. H. & Tsien, R. W. (1995). Functional impact of syntaxin on gating of N-type and Q-type calcium channels. *Nature*, 378, 623-626.

Braun, J. E. & Scheller, R. H. (1995). Cysteine string protein, a DnaJ family member, is present on diverse secretory vesicles. *Neuropharmacology*, 34, 1361-1369.

Braun, J. E., Wilbanks, S. M. & Scheller, R. H. (1996). The cysteine string secretory vesicle protein activates Hsc70 ATPase. *J Biol Chem*, 271, 25989-25993.

Cepeda, C., Hurst, R. S., Calvert, C. R., Hernandez-Echeagaray, E., Nguyen, O. K., Jocoy, E., Christian, L. J., Ariano, M. A. & Levine, M. S. (2003). Transient and progressive electrophysiological alterations in the corticostriatal pathway in a mouse model of Huntington's disease. *J Neurosci*, 23, 961-969.

Chamberlain, L. H. & Burgoyne, R. D. (1997). Activation of the ATPase activity of heat-shock proteins Hsc70/Hsp70 by cysteine-string protein. *Biochem J,* 322, 853-858.

Chandra, S., Gallardo, G., Fernandez-Chacon, R., Schluter, O. M. & Sudhof, T. C. (2005). Alpha-synuclein cooperates with CSPalpha in preventing neurodegeneration. *Cell*, 123, 383-396.

Cheetham, M. E. & Caplan, A. J. (1998). Structure, function and evolution of DnaJ: conservation and adaptation of chaperone function. *Cell Stress Chaperones*, 3, 28-36.

Chung, J., Nguyen, A. K., Henstridge, D. C., Holmes, A. G., Chan, M. H., Mesa, J. L., Lancaster, G. I., Southgate, R. J., Bruce, C. R., Duffy, S. J., Horvath, I., Mestril, R., Watt, M. J., Hooper, P. L., Kingwell, B. A., Vigh, L., Hevener, A. & Febbraio M. A. (2008). HSP72 protects against obesity-induced insulin resistance. *Proc Natl Acad Sci U S A*, 105, 1739–1744.

Ciechanover, A. & Brundin, P. (2003) The ubiquitin proteasome system in neurodegenerative diseases: sometimes the chicken, sometimes the egg. *Neuron*, 40, 427-446.

Colby, D. W., Chu, Y., Cassady, J. P., Duennwald, M., Zazulak, H., Webster, J. M., Messer, A., Lindquist, S., Ingram, V. M. & Wittrup, K. D. (2004). Potent inhibition of huntingtin aggregation and cytotoxicity by a disulfide bond-free single-domain intracellular antibody. *Proc Natl Acad Sci U S A,* 101, 17616-17621.

Connell, P., Ballinger, C. A., Jiang, J., Wu, Y., Thompson, L. J., Hohfeld, J. & Patterson, C. (2001). The co-chaperone CHIP regulates protein triage decisions mediated by heat-shock proteins. *Nat Cell Biol,* 3, 93-96.

Craig, E.A. (1985). The heat shock response. *CRC Crit Rev Biochem,* 18, 239-280.

Cummings, D. M., Andre, V. M., Uzgil, B. O., Gee, S. M., Fisher, Y. E., Cepeda, C. & Levine, M. S. (2009). Alterations in cortical excitation and inhibition in genetic mouse models of Huntington's disease. *J Neurosci,* 29, 10371-10386.

Cyr, D. M., Langer, T. & Douglas, M. G. (1994). DnaJ like proteins: molecular chaperones and specific regulators of hsp70. *TIBS*, 19, 176-181.

Dawson-Scully, K., Bronk, P., Atwood, H. L. & Zinsmaier, K. E. (2000). Cysteine-string protein increases the calcium sensitivity of neurotransmitter exocytosis in Drosophila. *J Neurosci*, 20, 6039-6047.

Dewhurst, K., Oliver, J. E. & McKnight, A. L. (1970). Socio-psychiatric consequences of Huntington's disease. *Br J Psychiatry*, 116, 255-258.

DiFiglia, M., Sapp, E., Chase, K. O., Davies, S. W., Bates, G. P., Vonsattel, J. P. & Aronin, N. (1997). Aggregation of huntingtin in neuronal intranuclear inclusions and dystrophic neurites in brain. *Science*, 277, 1990-1993.

Duff, K., Paulsen, J. S., Beglinger, L. J., Langbehn, D. R. & Stout, J. C. (2007). Psychiatric symptoms in Huntington's disease before diagnosis: the predict-HD study. *Biol Psychiatry,* 62, 1341-1346.

Dunah, A. W., Jeong, H., Griffin, A., Kim, Y. M., Standaert, D. G., Hersch, S. M., Mouradian, M. M., Young, A. B., Tanese, N. & Krainc, D. (2002). Sp1 and TAFII130 transcriptional activity disrupted in early Huntington's disease. *Science*, 296, 2238-2243.

Echtenkamp, F. J. & Freeman, B. C. (2011). Expanding the cellular molecular chaperone network through the ubiquitous cochaperones. *Biochim Biophys Acta,* 1823, 668-673.

Fernandes, H. B. & Raymond, L. A. (2009). NMDA Receptors and Huntington's Disease. In *Biology of the NMDA Receptor*, A. M. Van Dongen, ed. (Boca Raton (FL)).

Fernandez-Chacon, R., Wolfel, M., Nishimune, H., Tabares, L., Schmitz, F., Castellano-Munoz, M., Rosenmund, C., Montesinos, M. L., Sanes, J. R., Schneggenburger, R. & Südhof, T. C. (2004). The Synaptic Vesicle Protein CSPalpha Prevents Presynaptic Degeneration. *Neuron*, 42, 237-251.

Folstein, S., Abbott, M. H., Chase, G. A., Jensen, B. A. & Folstein, M. F. (1983). The association of affective disorder with Huntington's disease in a case series and in families. *Psychol Med,* 13, 537-542.

Fujikake, N., Nagai, Y., Popiel, H. A., Okamoto, Y., Yamaguchi, M. & Toda, T. (2008). Heat shock transcription factor 1-activating compounds suppress polyglutamine-induced neurodegeneration through induction of multiple molecular chaperones. *J Biol Chem,* 283, 26188-26197.

Gafni, J. & Ellerby, L. M. (2002). Calpain activation in Huntington's disease. *J Neurosci*, 22, 4842-4849.

Gafni, J., Hermel, E., Young, J. E., Wellington, C. L., Hayden, M. R. & Ellerby, L. M. (2004). Inhibition of calpain cleavage of huntingtin reduces toxicity: accumulation of calpain/caspase fragments in the nucleus. *J Biol Chem,* 279, 20211-20220.

Gething, M. & Sambrook, J. (1992). Protein folding in the cell. *Nature*, 355, 33-45.

Gil, J. M. & Rego, A. C. (2008). Mechanisms of neurodegeneration in Huntington's disease. *Eur J Neurosci*, 27, 2803-2820.

Gil, J. M. & Rego, A. C. (2009). The R6 lines of transgenic mice: a model for screening new therapies for Huntington's disease. *Brain Res Rev*, 59, 410-431.

Gil-Mohapel, J. M. (2011). Screening of Therapeutic Strategies for Huntington's Disease in YAC128 Transgenic Mice. *CNS Neurosci Ther*, [Electronic publication ahead of print].

Glover, J. R. & Lindquist, S. (1998). Hsp104, Hsp70, and Hsp40: a novel chaperone system that rescues previously aggregated proteins. *Cell*, 94, 73-82.

Goellner, G. M. & Rechsteiner, M. (2003). Are Huntington's and polyglutamine-based ataxias proteasome storage diseases? *Int J Biochem Cell Biol*, 35, 562-571.

Goldberg, Y. P., Nicholson, D. W., Rasper, D. M., Kalchman, M. A., Koide, H. B., Graham, R. K., Bromm, M., Kazemi-Esfarjani, P., Thornberry, N. A., Vaillancourt, J.P. & Hayden, M. R. (1996). Cleavage of huntingtin by apopain, a proapoptotic cysteine protease, is modulated by the polyglutamine tract. *Nat Genet,* 13, 442-449.

Gunawardena, S. & Goldstein, L. S. (2005). Polyglutamine diseases and transport problems: deadly traffic jams on neuronal highways. *Arch. Neurol.*, 62, 46-51.

Gundersen, C. B. & Umbach, J. A. (1992). Suppression cloning of the cDNA for a candidate subunit of a presynaptic calcium channel. *Neuron*, 9, 527-537.

Haldimann, P., Muriset, M., Vígh, L. & Goloubinoff, P. (2011). The novel hydroxylamine derivative NG-094 suppresses polyglutamine protein toxicity in Caenorhabditis elegans. *J Biol Chem,* 286, 18784-18794.

Hansson, O., Nylandsted, J., Castilho, R. F., Leist, M., Jäättelä, M. & Brundin, P. (2003). Overexpression of heat shock protein 70 in R6/2 Huntington's disease mice has only modest effects on disease progression. *Brain Res,* 970, 47-57.

Harjes, P. & Wanker, E. E. (2003). The hunt for huntingtin function: interaction partners tell many different stories. *Trends Biochem Sci,* 28, 425-433.

Hartl, F. U. (1996). Molecular chaperones in cellular protein folding. *Nature*, 381, 571-579.

Hay, D. G., Sathasivam, K., Tobaben, S., Stahl, B., Marber, M., Mestril, R., Mahal, A., Smith, D. L., Woodman, B. & Bates, G. P. (2004). Progressive decrease in chaperone protein levels in a mouse model of Huntington's disease and induction of stress proteins as a therapeutic approach. *Hum Mol Genet,* 13, 1389-1405.

Hendrick, J. P. & Hartl, F. U. (1993). Molecular chaperone functions of heat-shock proteins. *Annu Rev Biochem,* 62, 349-384.

Herbst, M. & Wanker, E. E. (2007). Small molecule inducers of heat-shock response reduce polyQ-mediated huntingtin aggregation. A possible therapeutic strategy. *Neurodegener Dis*, 4, 254-260.

Hermel, E., Gafni, J., Propp, S. S., Leavitt, B. R., Wellington, C. L., Young, J. E., Hackam, A. S., Logvinova, A. V., Peel, A. L., Chen, S. F., Hook, V., Singaraja, R., Krajewski, S., Goldsmith, P. C., Ellerby, H. M., Hayden, M. R., Bredesen, D. E. & Ellerby, L. M. (2004). Specific caspase interactions and amplification are involved in selective neuronal vulnerability in Huntington's disease. *Cell Death Differ*, 11, 424-438.

Hersch, S. M. & Ferrante, R. J. (2004) Translating therapies for Huntington's disease from genetic animal models to clinical trials. *NeuroRx*, 1, 298-306.

Jackrel, M. E. & Shorter, J. (2011). Shock and awe: unleashing the heat shock response to treat Huntington disease. *J Clin Invest,* 121, 2972-2975.

Jarvis, S. E., Barr, W., Feng, Z. P., Hamid, J. & Zamponi, G. W. (2002). Molecular determinants of syntaxin 1 modulation of N-type calcium channels. *J Biol Chem,* 277, 44399-44407.

Jarvis, S. E., Magga, J. M., Beedle, A. M., Braun, J. E. A. & Zamponi, G. W. (2000). G protein modulation of N-type calcium channels is facilitated by physical interactions between syntaxin 1A and Gbeta gamma. *J Biol Chem,* 275, 6388-6394.

Jarvis, S. E. & Zamponi, G. W. (2001). Distinct molecular determinants govern syntaxin 1A-mediated inactivation and G-protein inhibition of N-type calcium channels. *J Neurosci,* 21, 2939-2948.

Jensen, P., Sorensen, S. A., Fenger, K. & Bolwig, T. G. (1993). A study of psychiatric morbidity in patients with Huntington's disease, their relatives, and controls. Admissions to psychiatric hospitals in Denmark from 1969 to 1991. *Br J Psychiatry*, 163, 790-797.

Johnson, J. L. (2011). Evolution and function of diverse Hsp90 homologs and cochaperone proteins. *Biochim Biophys Acta*, 1823, 607-613.

Johnson, J. N., Ahrendt, E. & Braun, J. E. (2010). CSPalpha: the neuroprotective J protein. *Biochem Cell Biol,* 88, 157-165.

Kalmar, B., Novoselov, S., Gray, A., Cheetham, M. E., Margulis, B. & Greensmith, L. (2008). Late stage treatment with arimoclomol delays disease progression and prevents protein aggregation in the SOD1 mouse model of ALS. *J. Neurochem,* 107, 339–350.

Kaltenbach, L. S., Romero, E., Becklin, R. R., Chettier, R., Bell, R., Phansalkar, A., Strand, A., Torcassi, C., Savage, J., Hurlburt, A., Cha, G. H., Ukani, L., Chepanoske, C. L., Zhen, Y., Sahasrabudhe, S., Olson, J., Kurschner, C., Ellerby, L. M., Peltier, J. M., Botas, J. & Hughes, R. E. (2007). Huntingtin interacting proteins are genetic modifiers of neurodegeneration. *PLoS Genet,* 3, e82.

Kampinga, H. H., Kanon, B., Salomons, F. A., Kabakov, A. E. & Patterson C (2003). Overexpression of the cochaperone CHIP enhances Hsp70-dependent folding activity in mammalian cells. *Mol Cell Biol,* 23, 4948-4958.

Kieran, D., Kalmar, B., Dick, J. R., Riddoch-Contreras, J., Burnstock, G. & Greensmith, L. (2004). Treatment with arimoclomol, a coinducer of heat shock proteins, delays disease progression in ALS mice. *Nat Med,* 10, 402–405.

Kim, Y. J., Yi, Y., Sapp, E., Wang, Y., Cuiffo, B., Kegel, K. B., Qin, Z. H., Aronin, N. & DiFiglia, M. (2001). Caspase 3-cleaved N-terminal fragments of wild-type and mutant huntingtin are present in normal and Huntington's disease brains, associate with membranes, and undergo calpain-dependent proteolysis. *Proc Natl Acad Sci USA,* 98, 12784-12789.

Labbadia, J., Cunliffe, H., Weiss, A., Katsyuba, E., Sathasivam, K., Seredenina, T., Woodman, B., Moussaoui, S., Frentzel, S., Luthi-Carter, R., Paganetti, P. & Bates, G. P. (2011). Altered chromatin architecture underlies progressive impairment of the heat shock response in mouse models of Huntington disease. *J Clin Invest,* 121, 3306-3319.

Lecerf, J. M., Shirley, T. L., Zhu, Q., Kazantsev, A., Amersdorfer, P., Housman, D. E., Messer, A. & Huston, J. S. (2001). Human single-chain Fv intrabodies counteract in situ huntingtin aggregation in cellular models of Huntington's disease. *Proc Natl Acad Sci U S A*, 98, 4764-4769.

Leveque, C., el Far, O., Martin-Moutot, N., Sato, K., Kato, R., Takahashi, M. & Seagar, M. J. (1994). Purification of the N-type calcium channel associated with syntaxin and synaptotagmin. A complex implicated in synaptic vesicle exocytosis. *J Biol Chem*, 269, 6306-6312.

Li, H., Li, S. H., Johnston, H., Shelbourne, P. F. & Li, X.J. (2000). Amino-terminal fragments of mutant huntingtin show selective accumulation in striatal neurons and synaptic toxicity. *Nat Genet,* 25, 385-389.

Li, J. Y., Popovic, N. & Brundin, P. (2005). The use of the R6 transgenic mouse models of Huntington's disease in attempts to develop novel therapeutic strategies. *NeuroRx*, 2, 447-464.

Li, S. H., Cheng, A. L., Zhou, H., Lam, S., Rao, M., Li, H. & Li, X. J. (2002). Interaction of Huntington disease protein with transcriptional activator Sp1. *Mol Cell Biol,* 22, 1277-1287.

Li, S. H. & Li, X. J. (2004). Huntingtin-protein interactions and the pathogenesis of Huntington's disease. *Trends Genet,* 20, 146-154.

Magga, J. M., Jarvis, S. E., Arnot, M. I., Zamponi, G. W. & Braun, J. E. (2000). Cysteine string protein regulates G protein modulation of N-type calcium channels. *Neuron*, 28, 195-204.

Mangiarini, L., Sathasivam, K., Seller, M., Cozens, B., Harper, A., Hetherington, C., Lawton, M., Trottier, Y., Lehrach, H., Davies, S. W. & Bates, G.P. (1996). Exon 1 of the HD gene with an expanded CAG repeat is sufficient to cause a progressive neurological phenotype in transgenic mice. *Cell*, 87, 493-506.

Marsh, J. L., Pallos, J. & Thompson, L. M. (2003). Fly models of Huntington's disease. *Hum Mol Genet,* 2, R187-R193.

Martindale, D., Hackam, A., Wieczorek, A., Ellerby, L., Wellington, C., McCutcheon, K., Singaraja, R., Kazemi-Esfarjani, P., Devon, R., Kim, S.U., Bredesen, D. E., Tufaro, F. & Hayden, M. R. (1998). Length of huntingtin and its polyglutamine tract influences localization and frequency of intracellular aggregates. *Nat Genet,* 18, 150-154.

McLear, J. A., Lebrecht, D., Messer, A. & Wolfgang, W. J. (2008). Combinational approach of intrabody with enhanced Hsp70 expression addresses multiple pathologies in a fly model of Huntington's disease. *FASEB J*, 22, 2003-2011.

Meacham, G. C., Patterson, C., Zhang, W., Younger, J. M. & Cyr, D.M. (2001). The Hsc70 co-chaperone CHIP targets immature CFTR for proteasomal degradation. *Nat Cell Biol*, 3, 100-105.

Messer, A. & McLear, J. (2006). The therapeutic potential of intrabodies in neurologic disorders: focus on Huntington and Parkinson diseases. *BioDrugs*, 20, 327-333.

Miller, L. C., Swayne, L. A., Chen, L., Feng, Z. P., Wacker, J. L., Muchowski, P. J., Zamponi, G. W. & Braun, J. E. A. (2003a). Cysteine string protein (CSP) inhibition of N-type calcium channels is blocked by mutant huntingtin. *J Biol Chem,* 278, 53072-53081.

Miller, L. C., Swayne, L. A., Kay, J. G., Feng, Z. P., Jarvis, S. E., Zamponi, G. W. & Braun, J. E. A. (2003b). Molecular determinants of cysteine string protein modulation of N-type calcium channels. *J Cell Sci,* 116, 2967-2974.

Miller, T. W. & Messer, A. (2005a). Intrabody applications in neurological disorders: progress and future prospects. *Mol Ther,* 12, 394-401.

Miller, T. W., Zhou, C., Gines, S., MacDonald, M. E., Mazarakis, N. D., Bates, G. P., Huston, J. S. & Messer, A. (2005b). A human single-chain Fv intrabody preferentially targets amino-terminal Huntingtin's fragments in striatal models of Huntington's disease. *Neurobiol Dis,* 19, 47-56.

Miller, V. M., Nelson, R. F., Gouvion, C. M., Williams, A., Rodriguez-Lebron, E., Harper, S. Q., Davidson, B. L., Rebagliati, M. R. & Paulson, H. L. (2005c). CHIP suppresses polyglutamine aggregation and toxicity in vitro and in vivo. *J Neurosci,* 25, 9152-9161.

Murphy, R. C. & Messer, A. (2004). A single-chain Fv intrabody provides functional protection against the effects of mutant protein in an organotypic slice culture model of Huntington's disease. *Brain Res Mol Brain Res,* 121, 141-145.

Neuwald, A. F. & Hirano, T. (2000). HEAT repeats associated with condensins, cohesins, and other complexes involved in chromosome-related functions. *Genome Res,* 10, 1445-1452.

Nie, Z., Ranjan, R., Wenniger, J. J., Hong, S. N., Bronk, P. & Zinsmaier, K.E. (1999). Overexpression of cysteine-string proteins in Drosophila reveals interactions with syntaxin. *J Neurosci,* 19, 10270-10279.

Novoselova, T. V., Margulis, B. A., Novoselov, S. S., Sapozhnikov, A. M., van der Spuy, J., Cheetham, M. E. & Guzhova, I. V. (2005). Treatment with extracellular HSP70/HSC70 protein can reduce polyglutamine toxicity and aggregation. *J Neurochem,* 94, 597-606.

Pelham, H. R. (1986). Speculations on the functions of the major heat shock and glucose-regulated proteins. *Cell*, 46, 959-961.

Perrin, V., Régulier, E., Abbas-Terki, T., Hassig, R., Brouillet, E., Aebischer, P., Luthi-Carter, R. & Déglon, N. (2007). Neuroprotection by Hsp104 and Hsp27 in lentiviral-based rat models of Huntington's disease. *Mol Ther*, 15, 903-911.

Petersen, A., Gil, J., Maat-Schieman, M. L., Bjorkqvist, M., Tanila, H., Araujo, I. M., Smith, R., Popovic, N., Wierup, N., Norlen, P., Li, J. Y., Roos, R. A., Sundler, F., Mulder, H. & Brundin, P. (2005). Orexin loss in Huntington's disease. *Hum Mol Genet,* 14, 39-47.

Qin, Z. H., Wang, Y., Sapp, E., Cuiffo, B., Wanker, E., Hayden, M. R., Kegel, K. B., Aronin, N. & DiFiglia, M. (2004). Huntingtin bodies sequester vesicle-associated proteins by a polyproline-dependent interaction. *J Neurosci*, 24, 269-281.

Qiu, X. B., Shao, Y. M., Miao, S. & Wang, L. (2006). The diversity of the DnaJ/Hsp40 family, the crucial partners for Hsp70 chaperones. *Cell Mol Life Sci,* 63, 2560-2570.

Romero, E., Cha, G. H., Verstreken, P., Ly, C. V., Hughes, R. E., Bellen, H. J. & Botas, J. (2008). Suppression of neurodegeneration and increased neurotransmission caused by expanded full-length huntingtin accumulating in the cytoplasm. *Neuron*, 57, 27-40.

Rosas, H. D., Koroshetz, W. J., Chen, Y. I., Skeuse, C., Vangel, M., Cudkowicz, M. E., Caplan, K., Marek, K., Seidman, L. J., Makris, N., Jenkins, B. G. & Goldstein, J. M. (2003). Evidence for more widespread cerebral pathology in early HD: an MRI-based morphometric analysis. *Neurology*, 60, 1615-1620.

Rozas, J. L., Gomez-Sanchez, L., Tomas-Zapico, C., Lucas, J. J. & Fernandez-Chacon, R. (2010). Presynaptic dysfunction in Huntington's disease. *Biochem Soc Trans,* 38, 488-492.

Rozas, J. L., Gomez-Sanchez, L., Tomas-Zapico, C., Lucas, J. J. & Fernandez-Chacon, R. (2011). Increased neurotransmitter release at the neuromuscular junction in a mouse model of polyglutamine disease. *J Neurosci,* 31, 1106-1113.

Ruiz, R., Casanas, J. J., Sudhof, T. C. & Tabares, L. (2008). Cysteine string protein-alpha is essential for the high calcium sensitivity of exocytosis in a vertebrate synapse. *Eur J Neurosci,* 27, 3118-3131.

Samuni, Y., Ishii, H., Hyodo, F., Samuni, U., Krishna, M.C., Goldstein, S. & Mitchell, J. B. (2010). Reactive oxygen species mediate hepatotoxicity induced by the Hsp90 inhibitor geldanamycin and its analogs. *Free Radic Biol Med,* 48, 1559-1563.

Saudou, F., Finkbeiner, S., Devys, D. & Greenberg, M. E. (1998). Huntingtin acts in the nucleus to induce apoptosis but death does not correlate with the formation of intranuclear inclusions. *Cell*, 95, 55-66.

Schaffar, G., Breuer, P., Boteva, R., Behrends, C., Tzvetkov, N., Strippel, N., Sakahira, H., Siegers, K., Hayer-Hartl, M. & Hartl, F. U. (2004). Cellular toxicity of polyglutamine expansion proteins: mechanism of transcription factor deactivation. *Mol Cell,* 15, 95-105.

Schilling, G., Becher, M. W., Sharp, A.H., Jinnah, H. A., Duan, K., Kotzuk, J. A., Slunt, H. H., Ratovitski, T., Cooper, J. K., Jenkins, N. A., Copeland, N. G., Price, D. L., Ross, C.

A. & Borchelt, D. R. (1999). Intranuclear inclusions and neuritic aggregates in transgenic mice expressing a mutant N-terminal fragment of huntingtin. *Hum Mol Genet*, 8, 397-407.

Schmitz, F., Tabares, L., Khimich, D., Strenzke, N., de la Villa-Polo, P., Castellano-Munoz, M., Bulankina, A., Moser, T., Fernandez-Chacon, R. & Sudhof, T. C. (2006). CSPalpha-deficiency causes massive and rapid photoreceptor degeneration. *Proc Natl Acad Sci U S A,* 103, 2926-2931.

Sharma, M., Burre, J. & Sudhof, T. C. (2011). CSPalpha promotes SNARE-complex assembly by chaperoning SNAP-25 during synaptic activity. *Nature Cell Biol,* 13, 30-39.

Sheng, Z. H., Rettig, J., Cook, T. & Catterall, W. A. (1996). Calcium-dependent interaction of N-type calcium channels with the synaptic core complex. *Nature*, 379, 451-454.

Sittler, A., Lurz, R., Lueder, G., Priller, J., Lehrach, H., Hayer-Hartl, M. K., Hartl, F. U. & Wanker, E. E. (2001). Geldanamycin activates a heat shock response and inhibits huntingtin aggregation in a cell culture model of Huntington's disease. *Hum Mol Genet,* 10, 1307-1315.

Slow, E. J., Graham, R. K., Osmand, A. P., Devon, R. S., Lu, G., Deng, Y., Pearson, J., Vaid, K., Bissada, N., Wetzel, R., Leavitt, B. R. & Hayden, M. R. (2005). Absence of behavioral abnormalities and neurodegeneration in vivo despite widespread neuronal huntingtin inclusions. *Proc Natl Acad Sci U S A,* 102, 11402-11407.

Slow, E. J., van Raamsdonk, J., Rogers, D., Coleman, S. H., Graham, R. K., Deng, Y., Oh, R., Bissada, N., Hossain, S. M., Yang, Y. Z., Li, X. J., Simpson, E. M., Gutekunst, C. A., Leavitt, B. R. & Hayden, M. R. (2003). Selective striatal neuronal loss in a YAC128 mouse model of Huntington disease. *Hum Mol Genet,* 12, 1555-1567.

Smith, R., Brundin, P. & Li, J.Y. (2005). Synaptic dysfunction in Huntington's disease: a new perspective. *Cell Mol Life Sci,* 62, 1901-1912.

Smith, R., Klein, P., Koc-Schmitz, Y., Waldvogel, H. J., Faull, R. L., Brundin, P., Plomann, M. & Li, J. Y. (2007). Loss of SNAP-25 and rabphilin 3a in sensory-motor cortex in Huntington's disease. *J Neurochem,* 103, 115-123.

Southwell, A. L., Ko, J. & Patterson, P. H. (2009). Intrabody gene therapy ameliorates motor, cognitive, and neuropathological symptoms in multiple mouse models of Huntington's disease. *J Neurosci,* 29, 13589-13602.

Spargo, E., Everall, I. P. & Lantos, P. L. (1993). Neuronal loss in the hippocampus in Huntington's disease: a comparison with HIV infection. *J Neurol Neurosurg Psychiatry,* 56, 487-491.

Spradling, A., Penman, S. & Pardue, M. L. (1975). Analysis of drosophila mRNA by in situ hybridization: sequences transcribed in normal and heat shocked cultured cells. *Cell*, 4, 395-404.

Stanley, E. F. & Mirotznik, R. R. (1997). Cleavage of syntaxin prevents G-protein regulation of presynaptic calcium channels. *Nature*, 385, 340-343.

Steffan, J. S., Bodai, L., Pallos, J., Poelman, M., McCampbell, A., Apostol, B. L., Kazantsev, A., Schmidt, E., Zhu, Y. Z., Greenwald, M., Kurokawa, R., Housman, D. E., Jackson, G. R., Marsh, J. L. & Thompson, L. M. (2001). Histone deacetylase inhibitors arrest polyglutamine-dependent neurodegeneration in Drosophila. *Nature*, 413, 739-743.

Steffan, J. S., Kazantsev, A., Spasic-Boskovic, O., Greenwald, M., Zhu, Y. Z., Gohler, H., Wanker, E. E., Bates, G. P., Housman, D. E. & Thompson, L. M. (2000). The

Huntington's disease protein interacts with p53 and CREB-binding protein and represses transcription. *Proc Natl Acad Sci U S A,* 97, 6763-6768.

Stetler, R.A., Gan, Y., Zhang, W., Liou, A.K., Gao, Y., Cao, G. & Chen, J. (2010). Heat shock proteins: cellular and molecular mechanisms in the central nervous system. *Progress in neurobiology* 92, 184-211.

Swayne, L. A., Beck, K. E. & Braun, J. E. (2006). The cysteine string protein multimeric complex. *Biochem Biophys Res Commun*, 348, 83-91.

Swayne, L. A. & Braun, J. E. (2007). Aggregate-centered redistribution of proteins by mutant huntingtin. *Biochem Biophys Res Commun,* 354, 39-44.

Swayne, L. A., Chen, L., Hameed, S., Barr, W., Charlesworth, E., Colicos, M. A., Zamponi, G. W. & Braun, J. E. (2005). Crosstalk between huntingtin and syntaxin 1A regulates N-type calcium channels. *Mol Cell Neurosci,* 30, 339-351.

Tagawa, K., Marubuchi, S., Qi, M. L., Enokido, Y., Tamura, T., Inagaki, R., Murata, M., Kanazawa, I., Wanker, E. E. & Okazawa, H. (2007). The induction levels of heat shock protein 70 differentiate the vulnerabilities to mutant huntingtin among neuronal subtypes. *J Neurosci,* 27, 868-880.

Takano, H. & Gusella, J. (2002). The predominantly HEAT-like motif structure of huntingtin and its association and coincident nuclear entry with dorsal, an NF-kB/Rel/dorsal family transcription factor. *BMC Neuroscience* 3, 15.

The Huntington's Disease Collaborative Research Group. (1993). A novel gene containing a trinucleotide repeat that is expanded and unstable on Huntington's disease chromosomes. The Huntington's Disease Collaborative Research Group. *Cell*, 72, 971-983.

Tobaben, S., Thakur, P., Fernandez-Chacon, R., Sudhof, T. C., Rettig, J. & Stahl, B. (2001). A trimeric protein complex functions as a synaptic chaperone machine. *Neuron*, 31, 987-999.

Tobaben, S., Varoqueaux, F., Brose, N., Stahl, B. & Meyer, G. (2003). A brain-specific isoform of small glutamine-rich tetratricopeptide repeat-containing protein binds to Hsc70 and the cysteine string protein. *J Biol Chem,* 278, 38376-38383.

Vaccarino, A. L., Sills, T., Anderson, K. E., Bachoud-Levi, A. C., Borowsky, B., Craufurd, D., Duff, K., Giuliano, J., Groves, M., Guttman, M., Kupchak, P., Ho, A. K., Paulsen, J. S., Pedersen, K. F., Van Duijn, E., Van Kammen, D. P. & Evans, K. (2011). Assessment of depression, anxiety and apathy in prodromal and early Huntington disease. *PLoS Curr*, 3, RRN1242.

van Duijn, E., Kingma, E. M. & van der Mast, R. C. (2007). Psychopathology in verified Huntington's disease gene carriers. *J Neuropsychiatry Clin Neurosci,* 19, 441-448.

van Duijn, E., Reedeker, N., Giltay, E. J., Roos, R. A. & van der Mast, R. C. (2010). Correlates of apathy in Huntington's disease. *J Neuropsychiatry Clin Neurosci*, 22, 287-294.

Vígh L., Horváth I., Maresca B. & Harwood J. L. (2007). Can the stress protein response be controlled by 'membrane-lipid therapy'? *Trends Biochem Sci,* 32, 357–363.

Vonsattel, J. P. & DiFiglia, M. (1998). Huntington disease. *J Neuropathol Exp Neurol*, 57, 369-384.

Wellington, C. L., Ellerby, L. M., Gutekunst, C. A., Rogers, D., Warby, S., Graham, R. K., Loubser, O., van Raamsdonk, J., Singaraja, R., Yang, Y.Z., Gafni, J., Bredesen, D., Hersch, S. M., Leavitt, B. R., Roy, S., Nicholson, D. W. & Hayden, M. R. (2002).

Caspase cleavage of mutant huntingtin precedes neurodegeneration in Huntington's disease. *J Neurosci,* 22, 7862-7872.

Wellington, C. L., Ellerby, L. M., Hackam, A. S., Margolis, R. L., Trifiro, M. A., Singaraja, R., McCutcheon, K., Salvesen, G. S., Propp, S. S., Bromm, M., Rowland, K. J., Zhang, T., Rasper, D., Roy, S., Thornberry, N., Pinsky, L., Kakizuka, A., Ross, C. A., Nicholson, D. W., Bredesen, D. E. & Hayden, M. R. (1998). Caspase cleavage of gene products associated with triplet expansion disorders generates truncated fragments containing the polyglutamine tract. *J Biol Chem,* 273, 9158-9167.

Wellington, C. L., Singaraja, R., Ellerby, L., Savill, J., Roy, S., Leavitt, B., Cattaneo, E., Hackam, A., Sharp, A., Thornberry, N., Nicholson, D. W., Bredesen, D. E. & Hayden, M. R. (2000). Inhibiting caspase cleavage of huntingtin reduces toxicity and aggregate formation in neuronal and nonneuronal cells. *J Biol Chem,* 275, 19831-19838.

Wiser, O., Bennett, M. K. & Atlas, D. (1996). Functional interaction of syntaxin and SNAP-25 with voltage-sensitive L- and N-type Ca2+ channels. *EMBO J*, 15, 4100-4110.

Wolfgang, W. J., Miller, T. W., Webster, J. M., Huston, J. S., Thompson, L. M., Marsh, J. L. & Messer, A. (2005). Suppression of Huntington's disease pathology in Drosophila by human single-chain Fv antibodies. *Proc Natl Acad Sci U S A,* 102:11563-11568.

Wu, M. N., Fergestad, T., Lloyd, T. E., He, Y., Broadie, K. & Bellen, H. J. (1999). Syntaxin 1A interacts with multiple exocytic proteins to regulate neurotransmitter release in vivo. *Neuron*, 23, 593-605.

Wu, W. C., Wu, M. H., Chang, Y. C., Hsieh, M. C., Wu, H. J., Cheng, K. C., Lai, Y. H. & Kao, Y. H. (2010). Geldanamycin and its analog induce cytotoxicity in cultured human retinal pigment epithelial cells. *Exp Eye Res,* 91, 211-219.

Wyttenbach, A., Carmichael, J., Swartz, J., Furlong, R. A., Narain, Y., Rankin, J. & Rubinsztein, D. C. (2000). Effects of heat shock, heat shock protein 40 (HDJ-2), and proteasome inhibition on protein aggregation in cellular models of Huntington's disease. *Proc Natl Acad Sci U S A,* 97, 2898-2903.

Xu, F., Proft, J., Gibbs, S., Winkfein, B., Johnson, J. N., Syed, N. & Braun, J. E. (2010). Quercetin targets cysteine string protein (CSPalpha) and impairs synaptic transmission. *PLoS One*, 5, e11045.

Yamanaka, T., Miyazaki, H., Oyama, F., Kurosawa, M., Washizu, C., Doi, H. & Nukina, N. (2008). Mutant Huntingtin reduces HSP70 expression through the sequestration of NF-Y transcription factor. *EMBO J*, 27, 827-839.

Young, J. C., Agashe, V. R., Siegers, K. & Hartl, F. U. (2004). Pathways of chaperone-mediated protein folding in the cytosol. *Nat Rev Mol Cell Biol,* 5, 781-791.

Young, J. C., Barral, J. M. & Ulrich, H. F. (2003). More than folding: localized functions of cytosolic chaperones. *Trends Biochem Sci*, 28, 541-547.

Zeron, M. M., Hansson, O., Chen, N., Wellington, C. L., Leavitt, B. R., Brundin, P., Hayden, M. R. & Raymond, L. A. (2002). Increased sensitivity to N-methyl-D-aspartate receptor-mediated excitotoxicity in a mouse model of Huntington's disease. *Neuron*, 33, 849-860.

Zinsmaier, K. E. (2010). Cysteine-string protein's neuroprotective role. *J Neurogenet,* 24, 120-132.

Zinsmaier, K. E. & Bronk, P. (2001). Molecular chaperones and the regulation of neurotransmitter exocytosis. *Biochem Pharmacol*, 62, 1-11.

Zinsmaier, K. E., Eberle, K. K., Buchner, E., Walter, N. & Benzer, S. (1994). Paralysis and early death in cysteine string protein mutants of Drosophila. *Science*, 263, 977-980.

Zinsmaier, K. E., Hofbauer, A., Heimbeck, G., Pflugfelder, G. O., Buchner, S. & Buchner, E. (1990). A cysteine-string protein is expressed in retina and brain of Drosophila. *J Neurogenet*, 7, 15-29.

Zourlidou, A., Gidalevitz, T., Kristiansen, M., Landles, C., Woodman, B., Wells, D. J., Latchman, D. S., de Belleroche, J., Tabrizi, S. J., Morimoto, R. I. & Bates, G. P. (2007). Hsp27 overexpression in the R6/2 mouse model of Huntington's disease: chronic neurodegeneration does not induce Hsp27 activation. *Hum Mol Genet*, 16, 1078-1090.

In: Heat Shock Proteins
Editor: Saad Usmani
ISBN: 978-1-62417-571-8

Chapter V

Roles of Heat Shock Proteins of *Sesamia Nonagrioides* in Developmental Processes and Abiotic Stress Response

Anna Kourti*[*], *Theodoros Gkouvitsas and Dimitrios Kontogiannatos
Department of Agricultural Biotechnology,
Agricultural University of Athens, Athens, Greece

Abstract

This study aims to present available data regarding the implication of the invertebrate heat shock proteins in abiotic stress response and in other major physiological and developmental pathways that play a critical role in insect growth and differentiation. Studying the transcriptional regulation of several heat shock proteins in the moth *Sesamia nonagrioides* (Lepidoptera: Noctuidae), (two members of the a-crystallin/sHsp family, the *SnoHsp19.5* and the *SnoHsp20.8*; two members of the heat shock 70 family, the *SnoHsc70* and the *SnoHsp70*; the unique member of the Hsp90 family, the *SnoHsp83*), we showed that these proteins may play various roles in the developmental processes, regulating important physiological mechanisms, like diapause programming.

The heat shock protein genes respond differently to heat/cold stresses and diapause conditions in *S. nonagrioides*. Expression patterns of *SnoHsp19.5* and *SnoHsp20.8* in nondiapausing individuals under different environmental conditions (heat or cold) showed different accumulation profiles for the two genes after heat and cold treatment. *SnoHsp19.5* was consistently expressed, while *SnoHsp20.8* gene was down-regulated in deep diapause and was up-regulated at the termination of diapause, suggesting that these two genes play distinctive roles in the regulation of diapause. *SnoHsc*70 is constitutively expressed, and *SnoHsp*70 is heat-inducible in non-diapausing insects. *SnoHsp*70 is down regulated during diapause, while *SnoHsc*70 is induced as the larvae enter deep diapause.

[*] Department of Agricultural Biotechnology, Agricultural University of Athens, Iera Odos 75, 11855, Athens, Greece, Tel. 302105294615, e-mail: akourti@aua.gr.

High temperature stress during diapause has no further effect on transcript levels of *SnoHsc*70. Our results show that *SnoHsc*70 may play important roles in assisting protein conformation during specific stages of diapause. *SnoHsp83* is constitutively expressed in non-diapausing larvae and is induced 15-fold by heat. *SnoHsp83* displays a similar pattern to *SnoHsc70* under diapause conditions, when extra larval moults occur, indicating that could be involved in the developmental process that occurs between two moults.

Our results suggest that the heat shock protein genes in *S. nonagrioides* act not only as molecular chaperons *per se*, protecting the organism from several abiotic and environmental stresses, but could also play major roles in the developmental process, regulating important physiological mechanisms.

Keywords: *Sesamia nonagrioides*, Heat Shock proteins, diapause, abiotic stress response

Introduction

Heat shock proteins (Hsps) comprise a diverse group of different classes of proteins present in almost all forms of life and show transient increased expression in response to a rapid increase in temperature (Lindquist, 1986; Nover and Scharf, 1997). Many are present constitutively in cells, while others are induced when a cell or organism undergoes various types of environmental stresses such as heat, cold, desiccation or oxygen deprivation (Feder and Hofmann, 1999; Kregel, 2002). The Hsps function as molecular chaperones that protect cellular proteins during protein biosynthesis, including the recognition and binding of unfolded and non-native proteins (Parsell and Lindquist, 1993; Hart and Hayar-Hartl, 2002; Walter and Buchner, 2002).

The major Hsps can be assigned to families on the basis of their molecular weight: Hsp110, Hsp100, Hsp90, Hsp70, Hsp60, and small Hsp families (Gething, 1997). In insects there are four major Heat-shock gene families: the small *Hsp* family with molecular masses ranging from 20-30 kDa, the *Hsp*60 family with molecular masses of approximately 60 kDa, the *Hsp70* family with molecular masses of approximately 70 kDa and the *Hsp90* family with higher masses (Denlinger *et al.*, 2001).

The *Hsp70* family is one of the most highly conserved gene families and its proteins are the most widely studied (Gupta and Golding, 1993; Boorstein *et al.*, 1994). The *Hsp70* gene family contains both stress inducible and constitutively (*Hsc*70) expressed genes that share many common structural features. The primary structure of the *Hsp70* genes includes an amino-terminal ATP-binding domain $\approx$ 45 kDa and a carboxy-terminal substrate binding domain $\approx$ 25 kDa. Likewise, the inducible and constitutive Hsp70s have common regulatory features and domains (Hung *et al.*, 1998). Despite these similarities, the expression patterns of these proteins are quite different. The group of the *Hsp70* genes, is expressed at very low levels under normal conditions but is induced rapidly in responses to various stresses. On the contrary, the *Hsc70* gene is expressed in cells under normal conditions but does not change in response to stress such as heat or cold shock (Kiang and Tsokos, 1998). The expression pattern of many inducible and constitutive *Hsp70* genes is variable in response to different types and conditions of stressors and also in different stages during development (Rinehart *et al.*, 2000; Mahroof *et al.*, 2005).

The heat shock protein 90 (Hsp90) is an abundant molecular chaperone that is highly conserved from prokaryotes to eukaryotes. A unique feature of Hsp90 proteins is that they are constitutively expressed at substantial level during non-stress conditions (Arbona, 1993). In unstressed cells, Hsp90 can account for 1–2% of the total cytosolic protein (Picard, 2002). The up-regulation of Hsp90 in response to heat shock is not as robust as for other heat shock proteins. In most cases, a 10-15 fold increase in protein levels can be expected when comparing heat-stressed organisms to unstressed controls (Arbona, 1993). Hsp90 appears to have a specific function during recovery from stress, being especially important in the reactivation of stress-inactivated protein (Nathan *et al.*, 1997). Hsp90s are distinguished from other chaperones in that most of the known substrates are signal transduction proteins, the classical examples being steroid hormone receptors and signaling kinases (Picard *et al.*, 1990; Xu and Lindquist, 1993). Hsp90 interacts with steroid hormone receptors to facilitate receptor activation by steroids, and to prevent transcriptional activation in the absence of steroids (Craig *et al.*, 1993; Hendrick and Hartl, 1993). Moreover, Hsp90 in the eukaryotic cytosol interacts with a variety of co-chaperone proteins that assemble into a multichaperone complex (Young *et al.*, 2001). The highly conserved 25 kDa NH_2-terminal domain of Hsp90 is the binding site for ATP and geldamycin, a drug that specifically targets Hsp90 (Whitesell *et al.*, 1994). The conserved Hsp90 chaperone family includes the Hsp90 of the eykaryotic cytosol Hsp90α and β in humans, Hsp86 and Hsp82 in mice, Hsp83 in *Drosophila* and Hsc82 and Hsp82 in yeast (Young *et al.*, 2001). In *D. melanogaster* and most other insects, Hsp90 proteins are approximately 83 kDa in size (Blackman and Meselson, 1986). Unlike *Hsp90* in vertebrates, *Hsp90* genes for all insects studied so far exist as a single copy, except of *Anopheles albimanus*, which contains two (Benedict *et al.*, 1996). *Hsp90*genes have also been implicated in developmental regulation, such as morphological evolution and arrest of reproduction in *Drosophila* (Rutheford and Lindquist, 1998; Marcus, 2001), during the dauer stage of *Chaenorhabditis elegans* (Dalley and Golomb, 1992), and during the diapause of *Chillo suppressalis* and *Dellia_antiqua* (Sonoda *et al.*, 2006; Chen *et al.*, 2005). Studies in *D. triauraria* and the blowfly, *Lucillia sericata*, have not provided evidence of Hsp90 involvement in diapause (Goto and Kimura, 2004; Tachibana *et al.*, 2005), whereas in studies with *Sarcophaga crassipalpis Hsp90* transcripts were shown to be down-regulated during diapause (Rinehart and Denlinger, 2000). This discrepancy needs to be elucidated by using additional species and different types of diapauses.

Small heat shock proteins (sHsps) range in size from 12 to 43 kDa and play important role in the cellular defence of prokaryotic and eukaryotic organisms against a variety of internal and external stressors (Narberhaus, 2002). The sHsp family is having a similar domain to that of a-crystallin. This domain is perhaps responsible for the highly oligomeric structure of this family (Sun and MacRae, 2005; Stamler et al., 2005). Multimerization of sHsps is assumed to be crucial for their function as molecular chaperones ensuring correct folding, assembly and transport of newly synthesized polypeptides as well as removing abnormal cellular proteins. Therefore, increased expression of sHsps can extend an organism's tolerance to a variety of environmental insults such as heat, cold, salt, desiccation and oxidants (Feder and Hofmann, 1999; Hayward *et al.*, 2004). sHsps appear to be involved in various biological processes such as apoptosis besides protection against heat stress (MacRae, 2000; Arrigo and Landry, 1994; Leroux *et a*l., 1997). In insects, they are assumed to play a role in the regulation of diapause (Yocum *et al.*, 1998; Rinehart and Denlinger, 2000). sHsps have been studied in insects, including *Drosophila triaurana*, and *D.*

melanogaster (Goto *et al.*,1998; Goto and Kimura, 2004), the flesh fly, *Sarcophaga crassipalpis* (Yocum *et al.*, 1998; Rinehart *et al.*, 2000), the endoparasitic wasp, *Venturia canescens* (Reineke, 2005), the leaf beetle, the intertidal copepod *Tigriopus japonicus* (Seo *et al.*, 2006) *Gastrophysa atrocyanea* (Atungulu *et al.*, 2006), the Indian meal moth, *Plodia interpunctella* (Shirke *et al.*, 1998), the silkworm, *Bombyx mori* (Sakano *et al.*, 2006). The up-regulation of sHsps appears to be common to diapause in species representing diverse insect orders, including Diptera, Lepidoptera, Coleoptera, and Hymenoptera that occurs in different developmental stages (embryo, larva, pupa, adult) (Rinehart *et al.*, 2007). In contrast, recent studies propose that this is not the case in all insect species studied so far (Goto and Kimura, 2004; Tachibana *et al.*, 2005; Sonoda *et al.*, 2006). sHsps appear to be involved in various biological processes such as apoptosis besides protection against heat stress (MacRae, 2000; Arrigo and Landry, 1994; Leroux *et al.*, 1997). Multimerization of sHsps is assumed to be crucial for their function as molecular chaperones ensuring correct folding, assembly and transport of newly synthesized polypeptides as well as removing abnormal cellular proteins. Therefore, increased expression of sHsps can extend an organism's tolerance to a variety of environmental insults such as heat, cold, salt, desiccation and oxidants (Feder and Hofmann, 1999; Hayward *et al.*, 2004).

For most insects, short day length evokes a stage-specific developmental arrest, known as diapause. To be successful in a highly seasonal, temperature zone environment, insects must restrict growth and reproduction to a few months during the summer and survive the remainder of the year without feeding (Tauber *et al.*, 1986; Danks, 1987; Denlinger *et al.*, 2005). Insect diapause is a dynamic process consisting of several successive phases. Diapause is a more profound, endogenously and centrally mediated interruption that routes the developmental programme away from direct morphogenesis into an alternative diapause programme of succession of physiological events; the start of diapause usually precedes the advent of adverse conditions and the end of diapause need not coincide with the end of adversity (Danks, 1987; Kostal, 2006). The knowledge of the mechanism of diapause is essential for understanding the seasonal adaptation of insect species, and such information is also required for the development of effective pest management strategies; manipulating domesticated species used in pollination and silk production; developing accurate predictive models used to forecast periods of pest abundance; and increasing the shelf-life of parasitoids and predatory mites used in the biological control industry. Diapause also presents an interesting model for probing fundamental questions in development, and we are indebted to diapause studies for early insights into insect hormones. In addition, it can be argued that insect diapause may provide insights into questions on aging, obesity and disease transmission, and diapausing insects offer a potentially rich source of pharmaceutical agents that may contribute to improvement of human health.

Diapause represents an alternative developmental pathway prompted by unique patterns of gene expression. While many genes are down- regulated, a specific number are up-regulated (Flanagan *et al,* 1998). Among the genes that are up-regulated during diapause are several, but not all, genes that encode heat shock proteins (Rinechart *et al.*, 2000; Rinehart and Denlinger, 2000). Genes encoding certain stress proteins (Hsp23 and 70) are highly up-regulataed during diapause, while others are either unaffected (Hsc70) or are down-regulated (Hsp90) (Denlinger *et al.*, 2001). How Hsps may function in the long-term developmental arrest associated with diapause is unclear. Yocum *et al.* (1998) suggested that Hsps may persist for long periods during diapause and this is very interesting because extended expression of Hsps can lead to deleterious effects, including retardation and cessation of development (Feder *et al.*, 1992).

The stalk borer *Sesamia nonagrioides* (Lefebvre) (Lepidoptera: Noctuidae), is a multivoltine species, causes noticeable damage on maize boring galleries in the stem, throughout the Mediterranean basin. *S. nonagrioides* is one of the important pests of sorghum and corn in Mediterranean countries. This species overwinters as a mature larva and in Greece the first adults appear from early March to early May, in mild and cold areas, respectively (Fantinou *et al.*, 1995). During diapause larvae continue to feed, undergoing supernumerary moults (Fantinou et al., 1998). Photoperiod has been reported as the major factor controlling the induction and termination of larval diapause under laboratory conditions, while temperature could influence the response to day length (Hilal, 1977; Galighet, 1982; Eizaguirre *et al.*, 1994; Fantinou *et al.*, 1995).

The moth *Sesamia nonagrioides*: larva (6th instar), pupa and adult.

In our studies, the insects were obtained from an established laboratory colony of *S. nonagrioides,* derived from larvae collected in Kopais (latitude 38° 14′, Central Greece) in 2007, maintained at 25±1°C and reared on artificial diet (Tsitsipis *et. al.*, 1984). The colony of non-diapausing insects was reared under long day (LD) conditions (16:8, light: dark) at 25°C, while diapausing larvae were under short day (SD) conditions (10:14, light: dark) at 25°C. Larvae of *S. nonagrioides* under SD conditions underwent several to many extra larval molts and exhibit ~13 instars (Gkouvitsas *et al.*, 2008). The age of analysing larvae was measured in respect to the instar. In the following text, the ontogeny that includes diapause will be divided into thee main phase: (1) Pre-diapause, (2) Deep-diapause, (3) Post-diapause.

The expression of different heat shock proteins in the diapause of *S. nonagrioides* has been studied (Gkouvitsas *et al.*, 2009). *SnoHsp70* was down regulated during diapause, while *SnoHsc70* is induced as the larvae enter deep diapause. This provoked us to investigate whether other families of Hsps, such as Hsp90, are also up-regulated during diapause of *S. nonagrioides*. To answer this we studied the accumulation of *SnoHsp*83 mRNA under heat or cold stress in diapausing and non-diapausing larvae. *SnoHsp83* was constitutively expressed in non-diapausing and diapausing stages of corn borer and was induced from a low level to fifteen-fold by heat. *SnoHsp83* transcripts were induced as the larvae entered deep diapause, like *SnoHsc70*. The results indicated that the expression of *SnoHsp83* gene is a developmentally regulated component of the diapause program in *S. nonagrioides*. However, limited studies have reported on the expression of *Hsc70* during insect diapause. We determined the full *Hsc/Hsp*70 cDNA sequences and their deduced amino acid sequences of *S. nonagrioides*. In addition, we studied the accumulation of *Hsc/Hsp*70 mRNA under heat and cold stress in diapausing and non-diapausing larvae of *S. nonagrioides*. Our results revealed interesting differences in the expression patterns of the two genes during diapause, particularly regarding the induction of *Hsc70* and down regulation of *Hsp70*. Contrary to published results, we observed that *Hsc70* transcripts are induced as the larvae enter deep diapause, while high temperature stress during diapause has no further effect on transcript levels. Thus, *Hsc70* expression is a developmentally regulated component of the diapause program in *S. nonagrioides*. In order to unravel the potential contribution of *sHsps* transcripts to diapause of *S. nonagrioides*, we isolated two full-lengths cDNA sequences of *S. nonagrioides*, (*SnoHsp19.5* and *SnoHsp20.8*). The expression patterns showed that these genes are rapidly and highly induced after heat stress and suppressed after cold stress. We also observed the presence of the corresponding transcripts throughout diapause under normal or heat-shock conditions, indicating that the *SnoHsp19.5* and *SnoHsp20.8* genes may play distinctive roles in the regulation of the diapause process.

Isolation and Molecular Characterization of *S. Nonagrioides* Heat Shock Protein Genes

Hsc70 and *Hsp70* Genes

RACE-PCR was performed in order to obtain full-length *S. nonagrioides* cDNA fragments corresponding to *Hsc* and *Hsp70* genes respectively. The amplified fragments were isolated and the nucleotide sequences were determined. We obtained a 2166bp long sequence that had an open reading frame (ORF) spanning nucleotides 104-2065 (Figure 1A) and producing a putative protein of 653 amino acids with a deduced molecular weight of 71.5 kDa and putative pI 5.38. There was a 5'-untranslated region of 103 bp preceding the initiation codon (ATG) and a polyadenylation signal (AATAAA) at nucleotides 2134-2139, 12 nucleotides upstream from the poly (A) tail. Amino acid sequence comparisons revealed significant homology to several Heat shock cognate 70 genes from insects.. Therefore, this *S. nonagrioides* gene will be referred as *SnoHsc70*. The cDNA sequence was deposited in GenBank with accession number DQ004584.

The second full-length cDNA sequence (Figure 1B) was 2318 bp long and contained an open reading frame (ORF) spanning nucleotides 132-2090, producing a putative protein of 633 amino acids with a deduced molecular weight of 70.2 kDa and putative pI 5.76. There was a 5'-untranslated region of 131 bp preceding the initiation codon (ATG) and a polyadenylation signal (AATAAA) at nucleotides 2272-2277, 23 nucleotides upstream from the poly (A) tail. Amino acid sequence comparisons with both the NCBI GenBank and PROSITE database showed that this protein is an Hsp70 protein. Therefore, this *S.*

nonagrioides gene will be referred as *SnoHsp70*. The cDNA was deposited in GenBank with accession number EU430480.

```
tattgacttgggtactgtcgacctcgagtttttttttttttttttgccgagttactcta  61
cgagttaagtcaacgactgagatagttatagaaaatcaaaaaatggcagcaaaagccccc  121
                                              __ M  A  A  K  A  P    6
gctgtaggtattgacttgggtaccacttactcgtgcgtgggagttttccagcatggtaaa  181
 A  V  G  I  D  L  G  T  T  Y  S  C  V  G  V  F  Q  H  G  K   26
gtggagatcatcgcaaatgaccagggcaacaggaccacgccctcatatgtagcgttcacc  241
 V  E  I  I  A  N  D  Q  G  N  R  T  T  P  S  Y  V  A  F  T   46
gacaccgagcgtctcatcggagatgccgccaagaaccaggtggcgatgaacccaacaac  301
 D  T  E  R  L  I  G  D  A  A  K  N  Q  V  A  M  N  P  N  N   66
acaattttcgatgccaaacgtctcatcggacgcaaattcgaagatgctactgtacaagct  361
 T  I  F  D  A  K  R  L  I  G  R  K  F  E  D  A  T  V  Q  A   86
gacatgaagcactggcccttcgaggttgtcagtgatggtggcaagccaaagatcaaggtt  421
 D  M  K  H  W  P  F  E  V  V  S  D  G  G  K  P  K  I  K  V  106
gcatacaagggtgaagataaaaccttcttccctgaggaagttagctcaatggtgctcaca  481
 A  Y  K  G  E  D  K  T  F  F  P  E  E  V  S  S  M  V  L  T  126
aaaatgaaggaaactgccgaggcataccttggcaaaacggtgcagaatgcagtaatcaca  541
 K  M  K  E  T  A  E  A  Y  L  G  K  T  V  Q  N  A  V  I  T  146
gttccagcgtacttcaatgactcacagagacaagccacaaaagatgcaggtaccatctct  601
 V  P  A  Y  F  N  D  S  Q  R  Q  A  T  K  D  A  G  T  I  S  166
ggtttgaatgttctccgtattatcaatgaaccaactgctgctgcgattgcatacggcctt  661
 G  L  N  V  L  R  I  I  N  E  P  T  A  A  A  I  A  Y  G  L  186
gacaagaagggtagtggagaacgaaatgtactgatttttcgatctcggcggcggtacttt  721
 D  K  K  G  S  G  E  R  N  V  L  I  F  D  L  G  G  G  T  F  206
gatgtgtccatcctgaccatcgaggatggtatcttcgaagtaaagtccactgctggtgac  781
 D  V  S  I  L  T  I  E  D  G  I  F  E  V  K  S  T  A  G  D  226
acacatttgggaggagaggacttcgacaaccgcatggtcaaccactttgtgcaggagttc  841
 T  H  L  G  G  E  D  F  D  N  R  M  V  N  H  F  V  Q  E  F  246
aagaggaaatacaaaaaggaccttgctaccaacaagagggcccttaggcgattgcgcact  901
 K  R  K  Y  K  K  D  L  A  T  N  K  R  A  L  R  R  L  R  T  266
gcttgcgaaagggcgaagagaactctctcctcgtccacacaggctagcattgaaatcgac  961
 A  C  E  R  A  K  R  T  L  S  S  S  T  Q  A  S  I  E  I  D  286
tctctgttcgagggtatcgacttctacacgtccattaccagggctcgtttcgaggaactg 1021
 S  L  F  E  G  I  D  F  Y  T  S  I  T  R  A  R  F  E  E  L  306
aacgccgatctgtttagatccaccatggagcctgtggagaagtccctccgtgatgcgaag 1081
 N  A  D  L  F  R  S  T  M  E  P  V  E  K  S  L  R  D  A  K  326
atggacaaatctcaaatccacgatatcgtacttgtaggtggttctactcgtattcccaaa 1141
 M  D  K  S  Q  I  H  D  I  V  L  V  G  G  S  T  R  I  P  K  346
gtgcagaagctccttcaagacttctttaatggcaaggagcttaacaaatccatcaacccc 1201
 V  Q  K  L  L  Q  D  F  F  N  G  K  E  L  N  K  S  I  N  P  366
gacgaggccgtagcttatggtgccgccgtccaggccgccattttgcacggtgacaagtct 1261
 D  E  A  V  A  Y  G  A  A  V  Q  A  A  I  L  H  G  D  K  S  386
gaagaggtacaggatctgctgctgctcgacrtgactccgctgtcactcggtatcgaaacc 1321
 E  E  V  Q  D  L  L  L  L  D  X  T  P  L  S  L  G  I  E  T  406
gccggtggtgtcatgaccacccttatcaagcgcaacaccaccattcccaccaagcagact 1381
 A  G  G  V  M  T  T  L  I  K  R  N  T  T  I  P  T  K  Q  T  426
caaacctttaccacctgctctgacaaccagcctggagtactcattcaggtgttcgagggc 1441
 Q  T  F  T  T  C  S  D  N  Q  P  G  V  L  I  Q  V  F  E  G  446
gagcgcgccatgaccaaggataacaatttactcggaaagtttgagctgaccggcattcct 1501
 E  R  A  M  T  K  D  N  N  L  L  G  K  F  E  L  T  G  I  P  466
cccgcgccgcgtggcgtacctcaaatcgaagtcaccttcgacatcgacgctaacggcatt 1561
 P  A  P  R  G  V  P  Q  I  E  V  T  F  D  I  D  A  N  G  I  486
cttaacgtgtctgctgtcgagaaatcgactaacaaggagaacaagattaccatcaccaac 1621
 L  N  V  S  A  V  E  K  S  T  N  K  E  N  K  I  T  I  T  N  506
gacaagggccgtctttcaaaggaggagattgagcgcatggtcaacgaggctgagaaatac 1681
 D  K  G  R  L  S  K  E  E  I  E  R  M  V  N  E  A  E  K  Y  526
aggactgaggatgagaagcagaaggagacgatccaggctaagaacgctctggaatcttac 1741
 R  T  E  D  E  K  Q  K  E  T  I  Q  A  K  N  A  L  E  S  Y  546
tgcttcaacatgaagtccacaatggaggatgagaagctcaaggacaaaatctcagactct 1801
 C  F  N  M  K  S  T  M  E  D  E  K  L  K  D  K  I  S  D  S  566
gacaaacagactatcctggacaagtgcaacgacaccatcaaatggctggactccaatcgg 1861
 D  K  Q  T  I  L  D  K  C  N  D  T  I  K  W  L  D  S  N  R  586
ctggctgataaggaagaatacgagcacaagcagaaggagctggaaggaatctgcaaccct 1921
 L  A  D  K  E  E  Y  E  H  K  Q  K  E  L  E  G  I  C  N  P  606
attatcaccaagatgtaccagggagcaggtggtatgcccggcggtatgcccggtggcatg 1981
 I  I  T  K  M  Y  Q  G  A  G  G  M  P  G  G  M  P  G  G  M  626
cccggcttccctggcggcgctcccggcgctggaggcgctgcctccggtggcggtgccggc 2041
 P  G  F  P  G  G  A  P  G  A  G  G  A  A  S  G  G  G  A  G  646
cctaccatcgaagaggtcgattaaacattccaacagattatacagcatgcttttactgat 2101
 P  T  I  E  E  V  D  *                                        653
atttcagtattcaactacagtacagtatctccaataaatgaatgaaactgaaaaaaaaaa 2161
aaaaa                                                         2166
```

(A)

Figure 1. (continued).

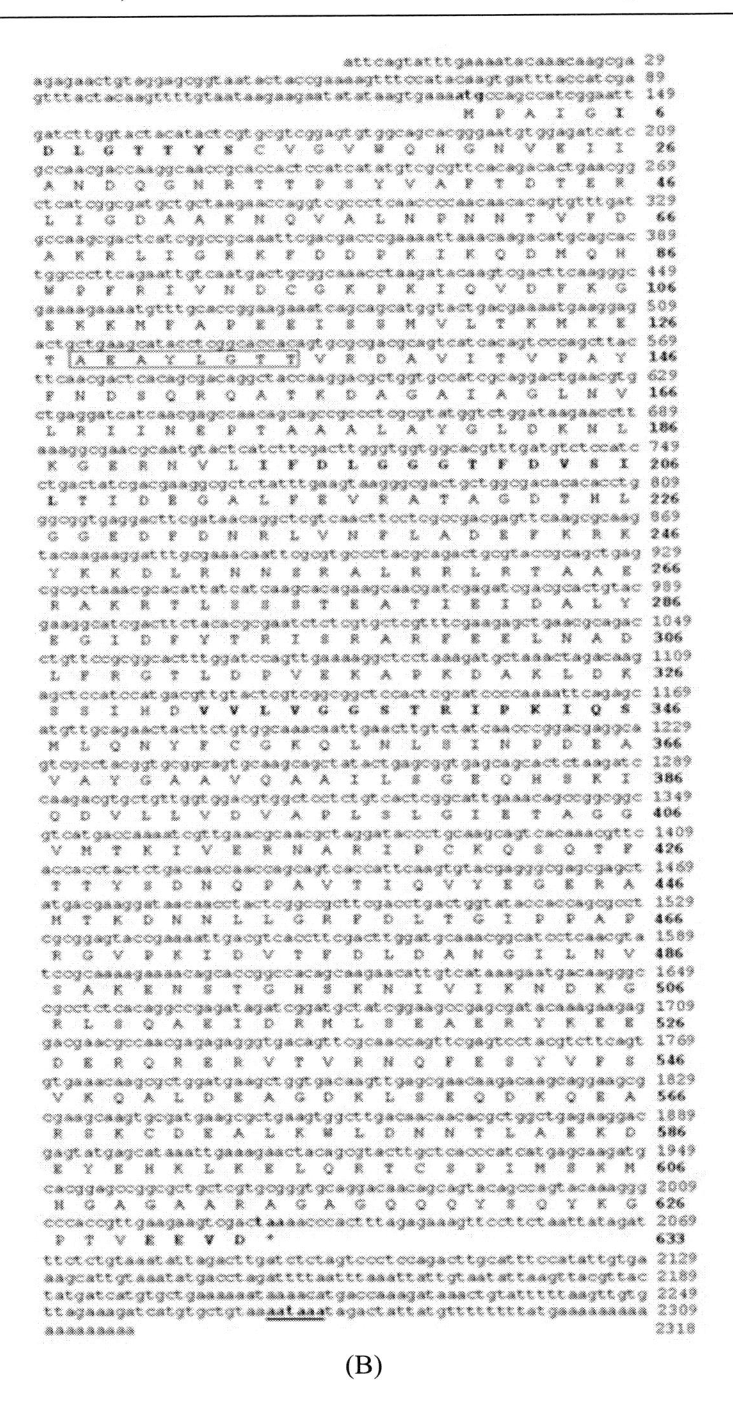

(B)

Figure 1. Nucleotide and deduced amino acid sequences of *S. nonagrioides Hsc70/Hsp70* cDNA. The deduced amino acid sequences are shown numbering begin at the initiation methionine. The consensus polyadenylation signal (AATAAA) is underlined and bold. 1. A (*SnoHsc70*). Three signatures at the positions 10-17, at 198-211, and at 335-349 are bolded; a putative ATP-GTP binding site at 132-139 is boxed; the cytoplasmic Hsp70 carboxyl terminal region at 650-653 is bolded. 1. B (*SnoHsp70*). Three signatures at the positions 6-13, at 194-207, and at 332-346 are bolded; a putative ATP-GTP binding site at 128-135 is boxed; the cytoplasmic Hsp70 carboxyl terminal region at 629-632 is bolded.

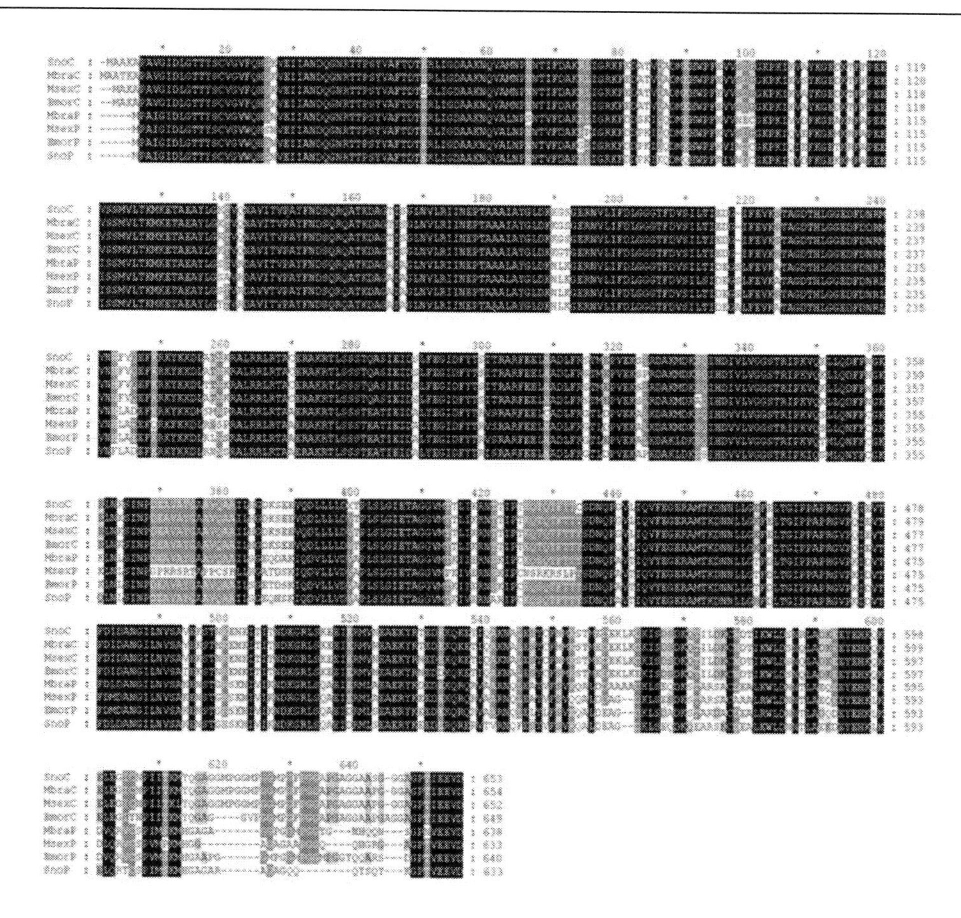

Figure 2. Multiple sequence alignment of the deduced amino acid sequence of *Hsc70/Hsp70* from *S. nonagrioides* with other cytosolic Hsc70s (with C as suffix) and Hsp70s (with P as suffix). Identical or similar amino acids are shaded black or grey, respectively. Abbreviations: SnoC = *S. nonagrioides* Hsc70; MbraC= *Mamestra brassicae* Hsc70; MsexC= *Manduca sexta* Hsc70; BmorC= *Bombyx mori* Hsc70; MbraP= *M. brassicae* Hsp70; MsexP= *M. sexta* Hsp70; BmorP= *B. mori* Hsp70; SnoP = *S. nonagrioides* Hsp70. Accession number of abbreviations are in Table 1.

In silico analysis of the SnoHsc/Hsp70 protein sequences revealed the presence of three Hsp70 family signature motifs.(Figure 1 A, B): IDLGTTYS at positions 10–17 and 6-13, respectively; IFDLGGGTFDVSIL at positions 198–211 and 194-207, respectively and IVLVGGSTRIPKVQK at positions 335–349 and VVLVGGSTRIPKIQS at positions 332–346. An ATP/GTP binding site motif (AEAYLGKT) was located in the N-terminus at amino acid positions 132–139 and AEAYLGTT at 128-135 (Figure1A). The COOH-terminal amino acid sequence Glu-Glu-Val-Asp (EEVD) was identified at the C-termini of SnoHsc/Hsp70 (Figure 1 A, B).

To retrieve protein sequences similar to those of SnoHsc/Hsp70, relevant databases were searched using the BLAST algorithm (Altschul *et al.*, 1990). The highest amino acid identity of SnoHsc70 (98%) was observed with the Hsc70 *of Manduca sexta*, *Bombyx mori* and *Mamestra brassicae* (Table 1). For SnoHsp70 the highest amino acid identity of 88% was observed with *B. mori* and 87% with *M. brassicae* and *Antheraea yamamai.* A comparison of the predicted amino acid sequence encoded by SnoHsc/Hsp70 with that of selected other insect Hsp70/Hsc70 family members is shown in Figure 2. According to Chen *et al.* (2006), in insects, Hsp70 class members were slightly shorter in length and smaller in molecular mass relatively to the Hsc70 class members. The length and mass differences are mainly due to two non-conservative regions, which are slightly shorter in Hsp70 proteins. The length of Hsp70 protein sequences range from 633 aa. (*S. nonagrioides*) to 640 aa. (*B. mori*), making them slightly shorter than those of Hsc70, that range from 649 aa. (*B. mori*) to 653 aa. (*S. nonagrioides*) (Figure 2).

To understand the evolutionary relationships of SnoHsc/Hsp70 with other related proteins, we conducted a phylogenetic analysis using proteins from different species as shown in Figure 3. To do this analysis, we obtained 13 protein sequences from databases. From these, 8 belong to Hsp70 and 5 belong to Hsc70 protein classes.

The tree was separated into four clades (Figure 3). Clades I and II are the Hsp70 clusters. Clade I includes the Lepidoptera, including SnoSsp70, and Clade II includes the Diptera. The other clades (III and IV) are the Hsc70 clusters, which clearly separated the lepidopterans, including SnoHsp70 (III) from the dipterans (IV). Our data provide further support that the Hsc/Hsp70 proteins presents good phylogenetic informativeness at taxonomic levels.

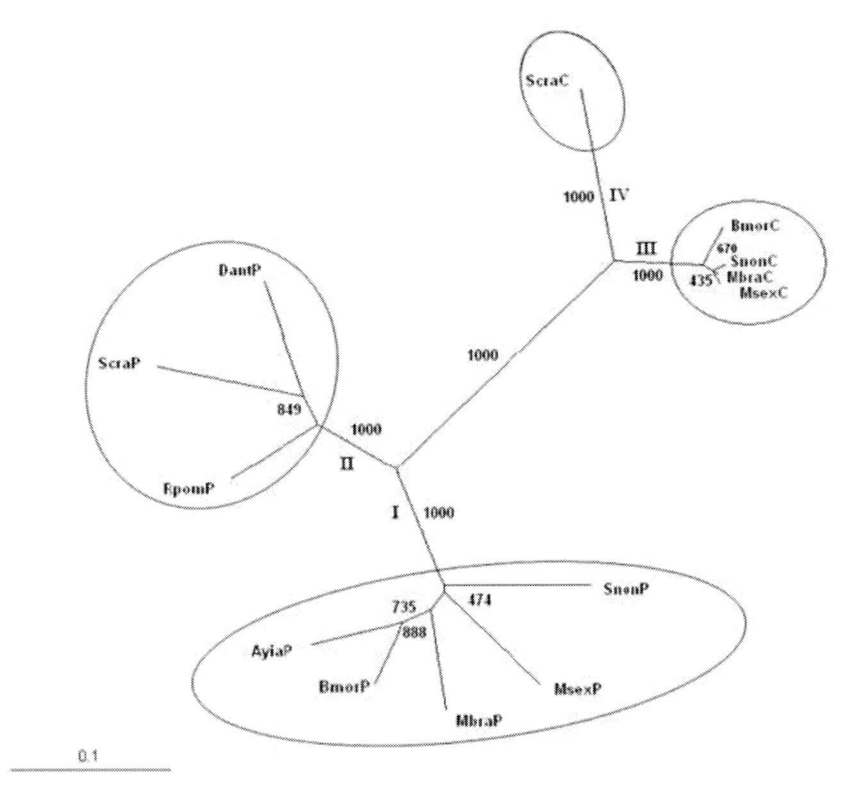

Figure 3. Phylogenetic Bootstrap Neighbour-Joining Tree (unrooted) built from CLUSTALX vl.8 default alignment for SnoHsc70/SnoHsp70 with the relevant proteins of other species. Bootstrap values for branches are shown (1000 replicates). The following abbreviations besides those described in the Table 2 are used: RpomP= *Rhagoletis pomonella* Hsp70 (Accession number EF103584); ScraP=*Sarcophaga crassipalpis* Hsp70 (AF107338); DantP= *Delia antiqua* Hsp70 (DQ017057); ScraP=*Sarcophaga crassipalpis* Hsc70 (Accession number AF107339).

Table 1. The homology of deduced amino acid sequences of the *S. nonagrioides Hsc70/Hsp70* genes with other members of *Hsc70/Hsp70* family

Species (genes)	*Hsc/Hsp70* family	Accession number	Amino identity (%)
Bombyx mori	*Hsc70*	BAB92074	98
Manduca sexta	*Hsc70*	AAF09496	98
Mamestra brassicae	*Hsc70*	BAF03556	98
Trichoplusia ni	*Hsc70*	AAB06239	97
Chilo suppressalis	*Hsc70*	BAE44308	97
Lonomia obliqua	*Hsc70*	AAV91465	97
Plutella xylostella	*Hsc70*	BAE48743	93
Bombyx mori	*Hsp70*	BAF69068	88
Mamestra brassicae	*Hsp70*	BAF03555	87
Plutella xylostella	*Hsp70*	BAF95560	85
Manduca sexta	*Hsp70*	AAO65964	84
Antheraea yamamai	*Hsp70*	BAD18974	87
Omphisa fuscidentalis	*Hsp70*	ABP93405	85
Rhagoletis pomonella	*Hsp70*	ABL06948	82

Hsp83 Gene

RT-PCR was performed in order to clone a cDNA fragment corresponding to *SnoHsp83*. A partial cDNa of 910 bp was isolated by RT-PCR using degenerate primers derived from *Hsp83* gene conserved domains.Based on the tag sequence, we designed 6 gene specific primers and amplified both the 5' and 3' ends of the cDNA. The full length of *Hsp83* cDNA of *S. nonagrioides*, assembled from the tag sequence and both 5' and 3' ends, is 2481 bp long and contains a unique open reading frame (ORF), which produced a putative protein of 717 amino acids with predicted molecular mass of 82.6 kDa. The start codon was located at position 173. The sequence around this methionine is very similar to the consensus *Drosophila* initiation sequence ([C/A] AA [A/C] AUG) (Cavener, 1987). There was a 5' untranslated region (5' UTR) of 173 bp and a complete 3' untranslated region (3' UTR) which contains a canonical poly(A) addition site (AATAAA) located 18 nucleotides upstream from the start of the poly(A) tract. The termination codon (TAA) occurred at nucleotide 2325. The cDNA and its deduced protein sequence were deposited in GenBank/EMBL/DDBJ with accession numbers DQ198859 and ABA54273 respectively. The complete nucleotide and deduced amino acid sequences of the *S. nonagrioides Hsp83* are shown in Figure 4.

All five highly conserved segments (motifs) defining the Hsp90 family signature of known eukaryotes (Gupta, 1995), were well conserved in the *S. nonagrioides* sequence (Figure 4). The terminal amino acid sequence MEEVD, constituting the core of the Hsp90 interaction surface for the tetratricopeptide repeats of Hsp90 co-chaperones (Ramsey *et al.*, 2000; Scheufler *et al.*, 2000) was strictly conserved and shared with the Hsp70 gene family. A search, via the BLAST software program (Altschul *et al.*, 1997), revealed strong homology between the deduced protein sequence and all members of the Hsp90 family, indicating that the isolated cDNA encodes a corn borer Hsp90 homologue, which was named SnoHsp83 following *Drosophila* nomenclature. Like all known cytosolic members of the Hsp90 family (Gupta, 1995), SnoHsp83 is comprised of two highly conserved domains, the N-terminal (aa

11–230) and middle (aa 259–615) domains, separated by a variable highly charged region (aa 231–258), and a less conserved C-terminal domain (aa 615–717).

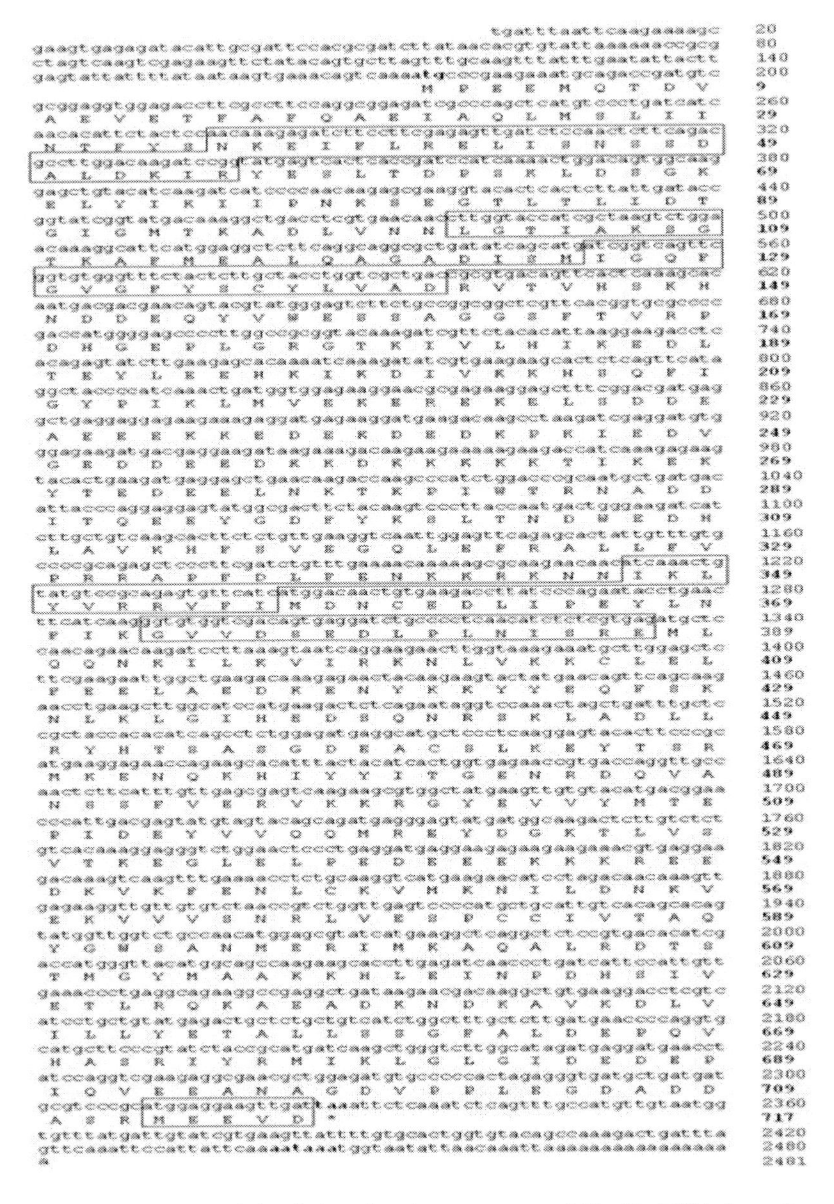

Figure 4. Nucleotide sequence and the deduced amino sequence of *Hsp83* cDNA from *S. nonagrioides* (*SnoHsp83*). The asterisk indicates the translational termination codon. A putative polyadenylation signal (AATAAA) is shown in bold letters. The five highly conserved amino acid segments that characterize all members of the Hsp90 family (Gupta, 1995) and the C-terminal pentapeptide MEEVD are shown boxed.

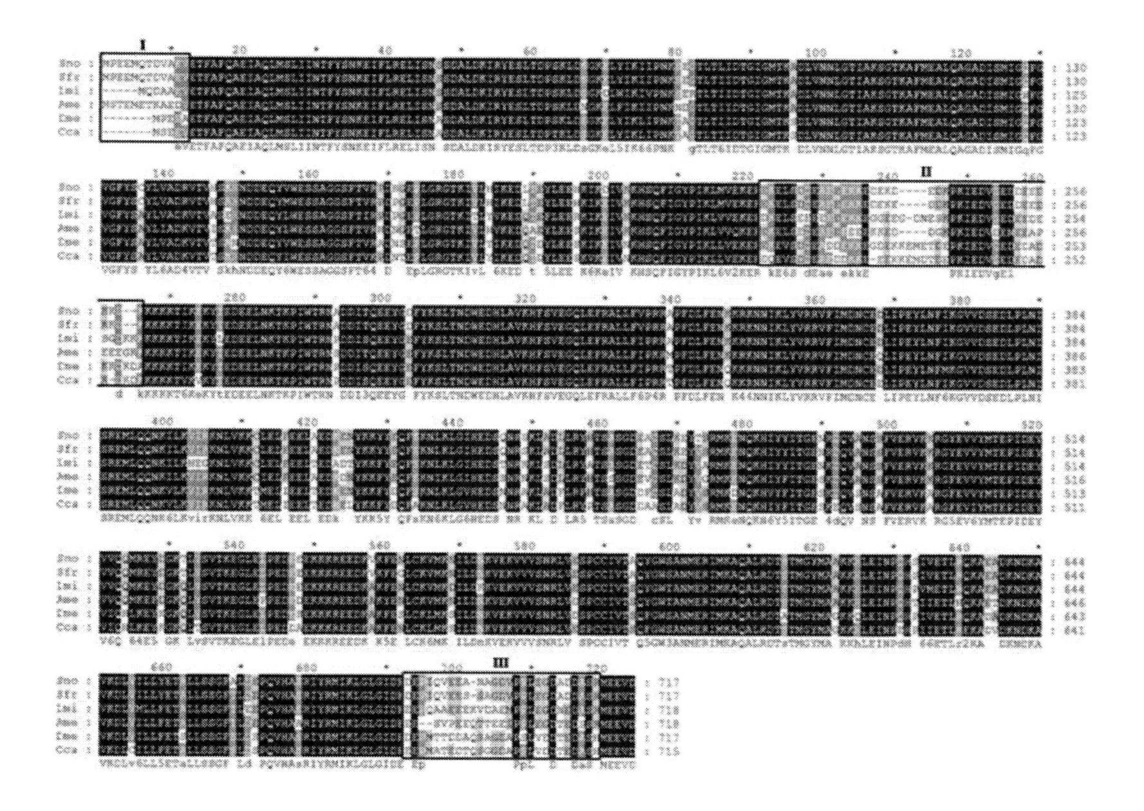

Figure 5. Alignment of *Sesamia nonagrioides* Hsp83 amino acid sequence with five other insect Hsp90 amino acid sequences. Identical or similar amino acids are shaded black or grey, respectively. Three non-conserved domains (I, II, III) of the Hsp90 protein are boxed. Abbreviations: Sno= *S. nonagrioides* Hsp83; Sfr= *S. frugiperda* Hsp83 (AAG44630); Dme= *Drosophila melanogaster* Hsp83 (NP523899); Cca = *Ceratitis capitata* Hsp83 (CAJ28987); (Lmi)= *Locusta migratoria* (AAS45246); (Ame)= *Apis mellifera* (XP623939).

To retrieve protein sequences similar to those of SnoHsp83, relevant databases were searched using the BLAST algorithm (Altschul *et al.*, 1997). The highest amino acid identity of SnoHsp83 (99% and 98%) was observed with the Hsp90 of *Mamestra brassicae* (BAF03554) and Hsp83 of *Spodoptera frugiperda* (AAG44630), respectively. We aligned the SnoHsp83 amino acid sequence with those of *S. frugiperda, Drosophila melanogaster, Ceratitis capitata, Locusta migratoria* and *Apis mellifera* (Figure 5). The result indicates that, despite the high conservation of the insect Hsp90 sequences, there are three amino acid blocks (shown boxed in Figure 5) which differ considerably among insect orders. The first (aa 1–12 in the SnoHsp83 sequence) is located at the N-terminus, the second (aa 220–260 in the SnoHsp83 sequence) includes the variable highly charged linker domain established by Gupta (1995) and the third (aa 686–712 in SnoHsp83 sequence) is located upstream to the conserved pentapeptide MEEVD.

In order to understand the evolutionary relationships of *SnoHsp83* with other related genes, we conducted a phylogenetic analysis using genes from different species as shown in Figure 6. The names of the species and the accession numbers of the sequences are given in the Figure legend. The results indicate that Lepidoptera are well segregated from Diptera,

Hemiptera, Hymenoptera and Orthoptera. Noctuidae, *S. nonagrioides, Mamestra brassicae* and *Spodoptera frugiperda,* were clustered in the same group and separated from the other lepidopteran. In lepidopteran insects such as *B. mori* and *S. frugiperda*, the *Hsp90* gene exists as a single copy in the genome (Landais *et al*., 2001). Similarly, our data from genomic Southern blot analyses also showed that *SnoHsp83* is encoded by a single gene in *S. nonagrioides* (Figure 7). Although in vertebrates there are two *Hsp90* genes displaying a high degree of sequence similarity that encode two cytosolic Hsp90 isoforms (alpha and beta)(Gupta, 1995), a single *Hsp90* gene has been reported for other insects studied to date, except of *Anopheles albimanus*, which contains two *Hsp82* genes with almost identical ORFs (Benedict *et al.,* 1996).

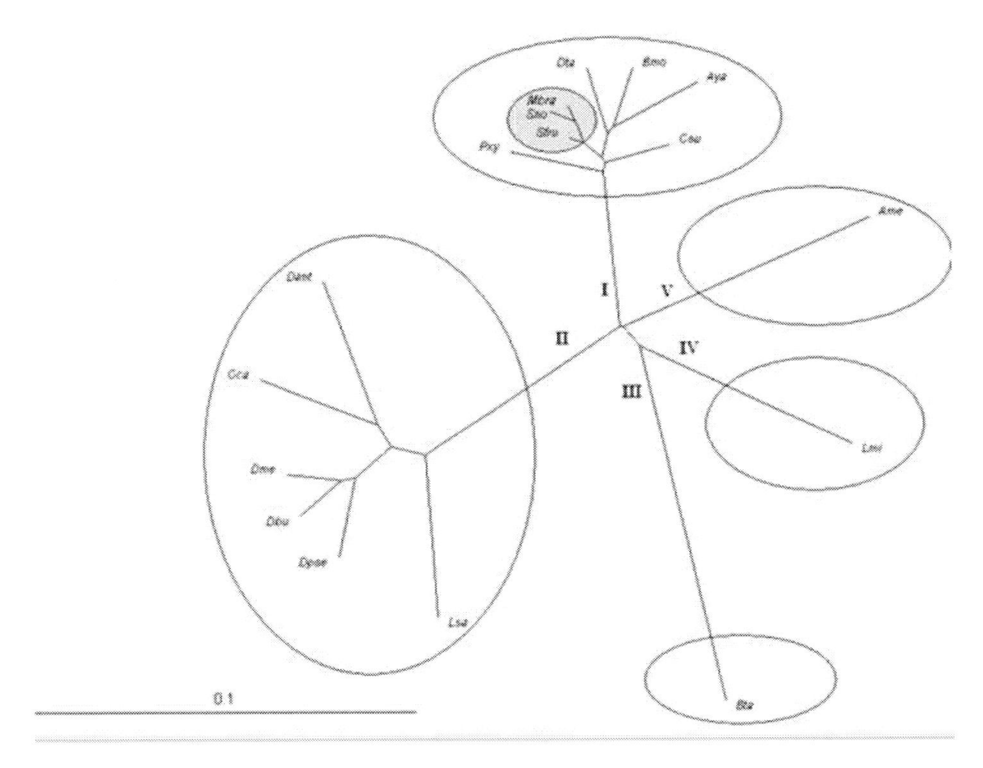

Figure 6. Phylogenetic analysis of insect Hsp90 sequences. An un-rooted neighbor-joining tree was constructed with Hsp90 amino acids sequences from 17 insect species. Bootstrap values (1000 replicates) are indicated on the nodes. The full names of species, the abbreviations and the accessions are: *Mamestra brassicae*)(Mbra); *Spodoptera frugiperda* (Sfr); *Chilo suppressalis* (Csu) (AB206477); *Bombyx mori* (Bmo); *Plutella xylostella* (Pxy) (BAE48742); *Antheraea yamamai* (Aya)(BAD15163); *Dendrolimus tabulaeformis* (Dta) (ABN09628); *Drosophila buzzatii* (Dbu) (ABK34943); *Liriomyza sativae* (Lsa) (AAW49253); *Ceratitis capitata* (Cca); *Drosophila pseudoobscura* (Dpse) (XP001353471); *Delia ant*iqua (Dant) (CAI64494); *Drosophila melanogaster* (Dme); *Locusta migratoria* (Lmi)(AAS45246); *Apis mellifera*(Ame)(XP623939), *Bemisia tabaci* (Bta) (AAZ17403).

Using restriction enzymes that do not have recognition sites within the probe, single hybridizing bands were observed. The *HindIII* digest gave one hybridizing band of 1.3 kb. *BclI* cuts 3 times in the *SnoHsp83* cDNA and resulted into a single hybridizing band of approximately 1.8 kb, as expected. These results indicate that the corn borer *Hsp83* gene exists in a single copy per haploid genome.

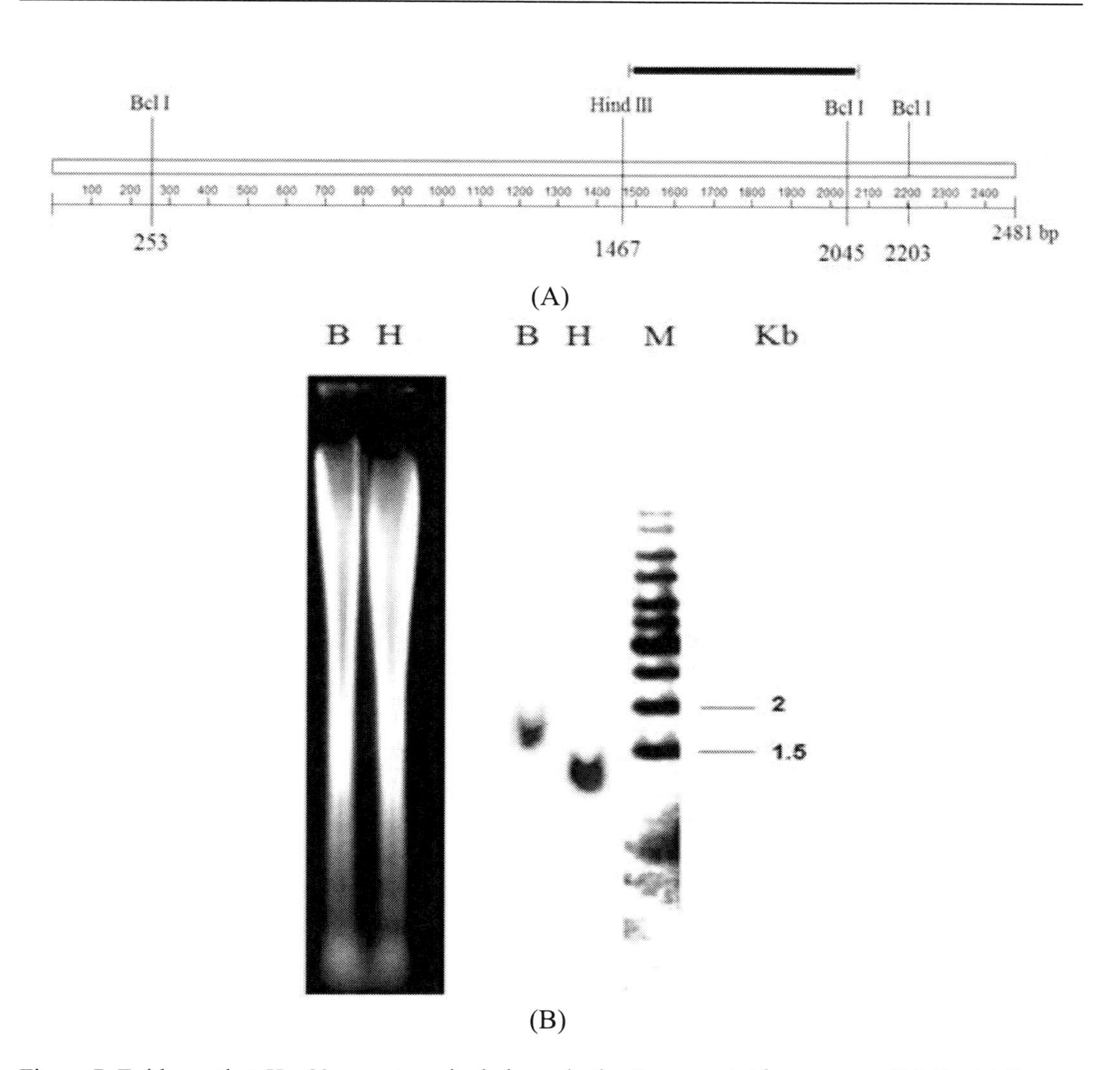

Figure 7. Evidence that *Hsp83* maps to a single locus in the *S. nonagrioides* genome. (A). Restriction map of corn stalk borer *Hsp83* cDNA. The bar above the cDNA indicates a 587 bp fragment used as a probe for genomic *Southern* blot analysis. (B). Genomic DNA from 5th instar larvae was digested with *Hind III* (H) and *BclI* (B) and subjected to Southern blot analysis, using as a probe the 587 bp fragment from the *Hsp83* cDNA shown above. DNA size markers are shown to the right of the autoradiogram.

Hsp19.5 and *Hsp20.8* Genes

The lengths of cDNA for *Hsp19.5* and *Hsp20.8* were 787 and 768 nucleotides, respectively (Figure 8). *Hsp19.5* cDNA (Figure 8.A) contained an open-reading frame of 522 bp, produced a putative protein of 174 amino acids with 5.97 pI and 19.5 kDa. There is a 5'-untranslated region of 144 bp preceding the initiation codon (ATG). The polyadenylation signal (AATAAA) is present at nucleotide 740. This gene was registered in GenBank with accession number EU668902.

Amino acid sequence comparisons with both the NCBI GenBank and PROSITE database showed that this protein has a high similarity to the small heat shock proteins (Hsp20) family.

Therefore, this *S. nonagrioides* gene will be referred as *SnoHsp19.5*. Through PROSITE analysis (Falquet *et al.*, 2002), we also found that SnoHsp19.5 has a secondary modification motifs, such as casein kinase II phosphorylation site (Figure 8A). Using the GOR4 program (http;//npsa-pbil.ibcp.fr/cgi-bin) (Garnier *et al.*, 1996), the secondary structures of the SnoHsp19.5 was predicted on the corresponding deduced amino-acid sequences and showed that 14.94% has an extended strand, 27.01% alpha helix region and 58.05% the random coil conformation. The cDNA of *Hsp20.8* (Figure 8. B) contained an open-reading frame of 555 bp, produced a putative protein of 185 amino acids with 6.24 pI and 20.8 kDa (Figure 8. B). The results indicated that the 5-transcript's end was mapping 93 bp upstream of the translation start site (ATG). The polyadenylation signal (AATAAA) is located at nucleotide 741. This gene was registered under DQ336356 in GenBank. Amino acid sequence comparisons with both the NCBI GenBank and PROSITE database showed that this protein has a high similarity to the α-crystallin-type small heat shock proteins (sHsp). Therefore, this *S. nonagrioides* gene will be referred as *SnoHsp20.8*. The putative α-crystallin domain composed of 84 residues in *Sno20.8*, is located between positions 62 and145 (Figure 8B). Through PROSITE analysis we also found that SnoHsp20.8 has several secondary modification motifs, such as an N-glycosylation site and two casein kinase II phosphorylation sites (Figure 8B). Using the GORA4 program, the secondary structures of the SnoHsp20.8 was predicted on the corresponding deduced amino-acid sequences and showed that 14.59% has an extended strand, 29.73% alpha helix region and 55.68 % the random coil conformation.

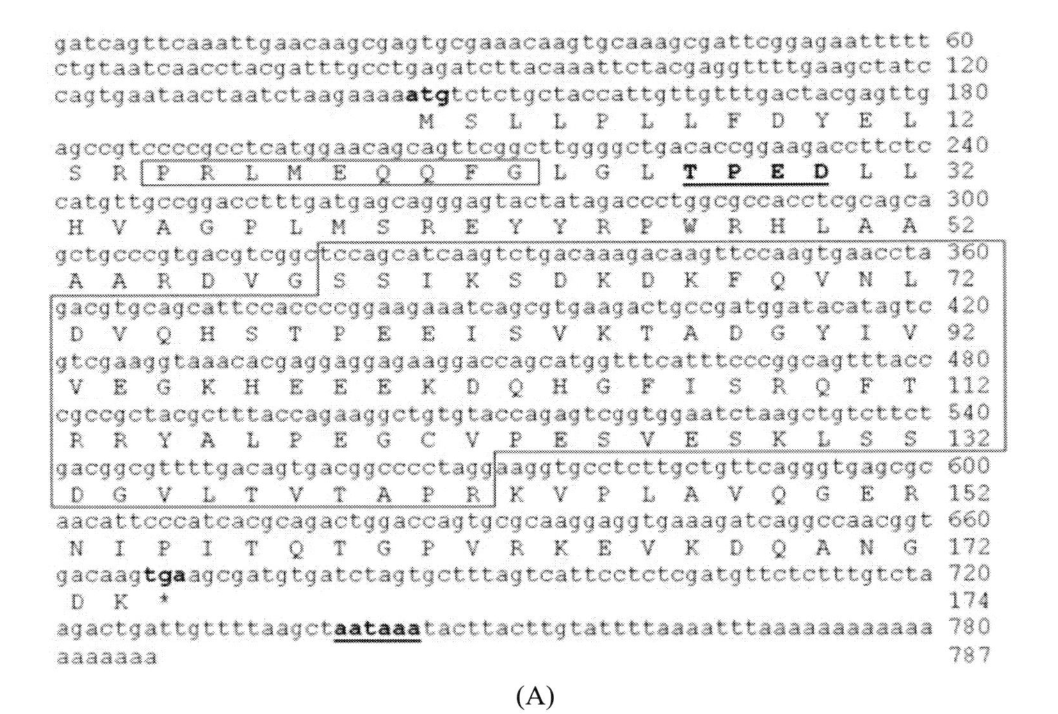

(A)

Figure 8. (Continued).

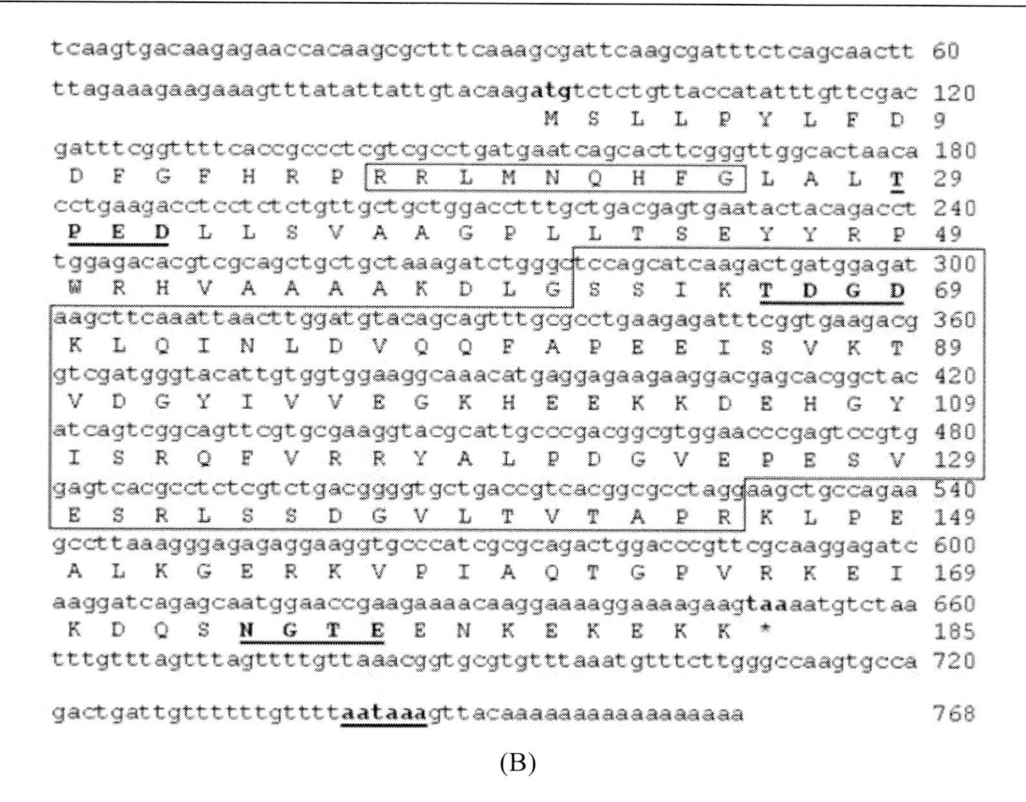

Figure 8. Nucleotide and deduced amino acid sequences of the cDNAs encoding *S. nonagrioides Hsp19.5* (A) and *Hsp20.8* (B). The predicted amino acid sequences of the 174 and 185 codons open reading frames are shown under the nucleotide sequences in the single-letter amino acid code. atg =start codon, tga= termination codon (A) and taa = termination codon (B), aataaa= polyadenylation signal. The boxed sections are the alpfa-crystallin domains. N glycosylation site (174-177 NGTE) (Figure 8B) and casein kinase II phosphorylation sites (27-30 TPED, 8A; 29-32 TPED, 8B and 66-69 TDGD, 8B) are underlined.

Table 2. Sequence identity between *SnoHsp19.5* and *SnoHsp20.8* deduced amino acid and several other insect sHsp proteins

Sequence name No	Insect	GenBank acces.	Sequence *Hsp19.5*	Identity (%) *Hsp20.8*
Hsp19.5	*S. nonagrioides*	EU668902	-	75
hsp20.8	*S. nonagrioides*	AB118968	75	-
hsp20.4	*Bombyx mori*	AF315318	79	83
hsp19.9	*B. Mori*	BAD74195	*84*	79
hsp19.7	*Chilo suppressalis*	BAE94664	*85*	78
hsp19.7	*Mamestra brassicae*	BAF03558	90	77
Hsp2	*Lonomia oblique*	AAV91361	75	74
hsp20.8	*B. mori*	AF315317	71	72
hsp21.5	*Choristoneura fumiferana*	AAZ14790	69	70
acr25	*Plodia interpunctella*	AAC36146	68	69
hsp19.8	*Choristoneura fumiferana*	AAZ14792	73	67

To retrieve protein sequences similar to those of *SnoHsp19.5* and *SnoHsp20.8*, relevant databases were searched using the BLAST algorithm (Altschul et al., 1990). The deduced amino acid sequence of SnoHsp19.5 and SnoHsp20.8 displayed a high degree of homology with those of other insects so far reported (Table 2). For SnoHsp19.5 the highest amino acid identity of 90% was observed with the Hsp19.7 protein of *Mamestra brassicae*; SnoHsp20.8 has 83% identity with the Hsp20.4 protein of *Bombyx mori.* The two proteins, SnoHsp19.5 and SnoHsp20.8 have 75 % identity.

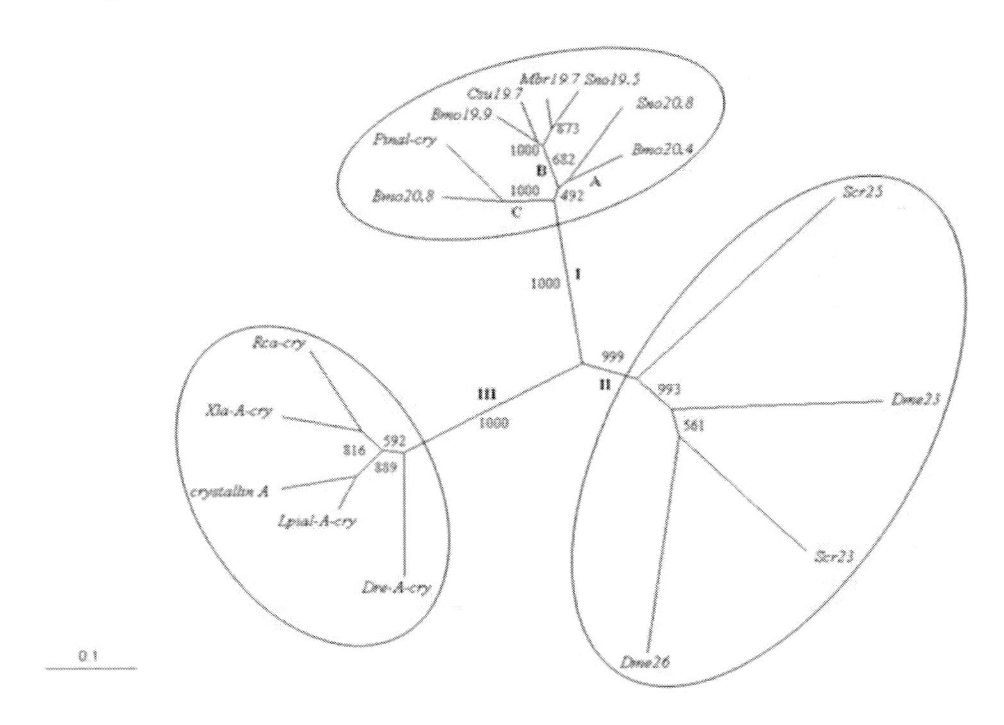

Figure 9. Phylogenetic Bootstrap Neighbour-Joining Tree (unrooted) built from CLUSTALX vl.8 default alignment for *Sno19.5* and *Sno20.8* with known sHsp protein sequences. Bootstrap values for braches are shown (1000 replicates). GenBank accession number and abbreviations (in parenthesis) for these sequences are: *Mamestra brassicae*: heat shock protein 19.7, BAF03558 (*Mbr19.7*), *Bombyx mori*: heat shock protein 20.8, AF315317 (*Bmo20.8*), *Plodia interpunctella*: alpha-crystallin cognate protein 25, AAC36146 (*Pinal-cry*), *Sarcophaga crassipalpis*: 25 kDa small heat shock of AB206941 (*Scr25*); 23 kDa small heat shock, AAC63387 (*Scr23*); *Drosophyla melanogaster*: 23 kDa heat shock protein, 523999 (*Dme23*), 26 kDa heat shock protein, CAA27526 (*Dme26*), *Rana catesbeiana*: A-crystallin a-chain, Q91311 (*Rca-cry*), *Xenopous laevis:* alpha A crystalline, BAA76897 (*Xla-A-cry*), *Latus ridibunds*: crystalline alpha A ,1010303 *(crystalline A*), *Lygodactylus picturatus*: alpha A-crystallin, CAFO2103 (*Lpial-A-cry*), *Danio rerio*: crystalline alpha A , NP6944482 (*Dre-A-cry*).

This comparison also showed that all of them had an α-crystallin domain and an N-terminal motif (James *et al.,* 1994). The N-terminal hydrophobic aggregation site was present in SnoHsp19.5 and SnoHsp20.8 (Figure 8.A,B) as has been found in other lepidopterans. In *S. nonagrioides*, the number of Cys residues differentiated the two sHsps from each other, as in sHsps of *B. mori* (Sakano *et al.*, 2006). SnoHsp19.5 had *0.6% Cys,* while SnoHsp20.8 had no Cys residues. To expose the relationship of *SnoHsp19.5* and *SnoHsp20.8* with other related genes, we conducted a phylogenetic analysis by collecting amino acid sequences from various sHsps and crystallins. As shown in Figure 9, the phylogenetic tree was separated into three clades. Clades I and II consisted of the *Hsp20* and *Hsp27* gene families, respectively, while clade III contained crystallin genes. Phylogenetic analysis indicated clade I, that consisted of the group of *Hsp20* gene family, separated from the other groups. This group was divided into three branches: the one (A) containing the *SnoHsp20.8* and the *B. mori Hsp20.4*, the second (B) containing the *B. mori Hsp19.9* and *SnoHsp19.5,* and the third (C) containing the *B. mori Hsp20.8.*

Heat Shock Protein Gene Expression and Abiotic Stress

Heat Shock Protein Gene Expression during Heat or Cold Stress

SnoHsc/Hsp70 Genes

To investigate whether different types of environmental stresses such as high or low temperature affect *SnoHsc/Hsp70* gene expression, diapausing and non-diapausing larvae of *S. nonagrioides* were exposed to various temperatures. The *SnoHsc/Hsp70* gene transcripts were amplified by semi-quantitative RT-PCR and quantified using Real-Time PCR. Transcript levels of *SnoHsc70* in non-diapausing larvae increased after the onset of heat shock (Figure 10. A, B). A 40°C heat shock for 45 minutes resulted in a 2-2.5 fold increase of transcript abundance. This increase was apparent when insects were incubated for 15 min to heat stress. Longer exposure to stress had almost no additional effect indicating that the *SnoHsc70* gene reached almost maximum levels of expression within 15 min.

A 40°C heat shock for 45 minutes resulted in a 130-fold increase of *SnoHsp70* transcript abundance. There was a 10-fold increase in mRNA accumulation within 15 min after exposure to heat stress and a 30-fold increase after 30 min. Longer exposure to stress (45 min) reached the maximum level of gene induction. These results clearly show that this gene is a bona fide heat-shock gene. There was an apparent and rapid gene induction within a few minutes after heat shock.

Even though exposure of larvae to low temperatures had a moderate affect on the expression of *SnoHsc70, SnoHsp70* expression changed radically (Figure 11). All lower temperatures suppressed expression of *SnoHsp70*. However, when larvae were let to recover at normal temperatures, there was an apparent induction when the larvae were exposed at rather low temperatures (-5 or 0°C).

Therefore induction was apparent when the absolute value difference of the temperature was high enough, indicating that this gene is highly heat inducible. Cold treatments did not affect expression of *SnoHsc70*. These results indicated that the two genes respond differently when larvae were stressed. *SnoHsp*70 transcripts are usually undetectable or constitutively expressed at low levels under normal (non-diapausing) conditions, but transcription is induced by heat stress in a variety of insect species (Goto *et al.*, 1998; Rinehart *et al.*, 2000; Tachibana *et al.*, 2005).

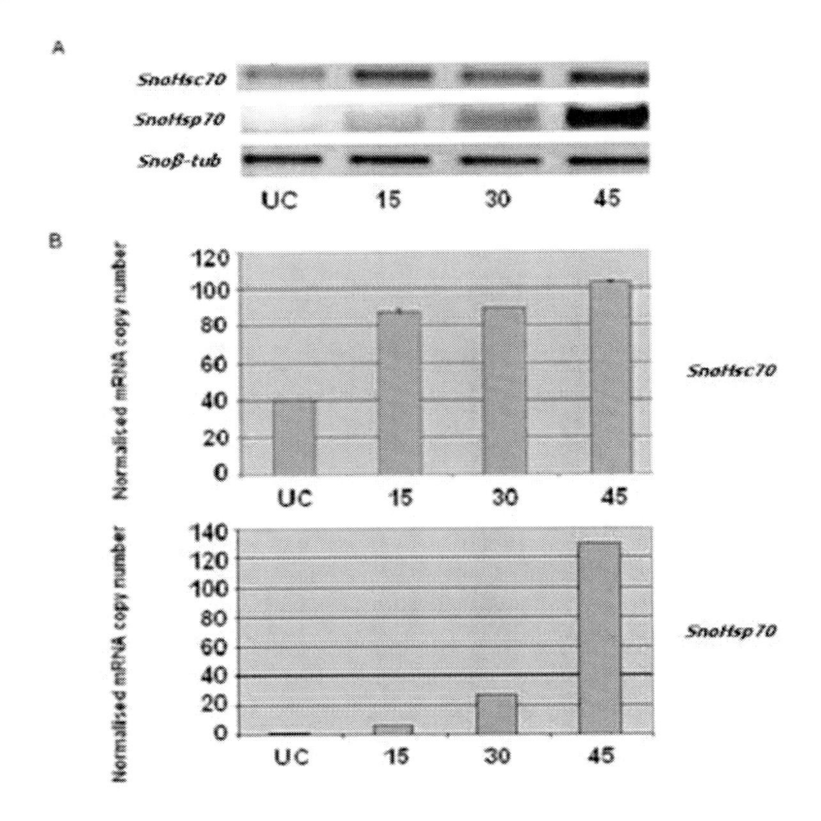

Figure 10. A. Semiquantitative RT-PCR expression analysis of *SnoHsc70/Hsp70* in response to heat shock in non-diapausing larvae (5th instar-25 days since hatching), exposed to 40 °C for 15, 30, and 45 min. UC: untreated control. *Sno*β-tubulin was used as the control gene. B. Real-Time PCR analysis of *SnoHsc70/Hsp70* genes in response to heat shock in non-diapausing larvae (5th instar-25 days since hatching), exposed to 40 °C for 15, 30, and 45 min. UC: untreated control. The relative mRNA copy number of the *SnoHsc70/Hsp70* genes transcripts are shown in relation to the number of *Sno*β-tubulin transcripts for each cDNA sample. The error bars refer to the standard error of the mean.

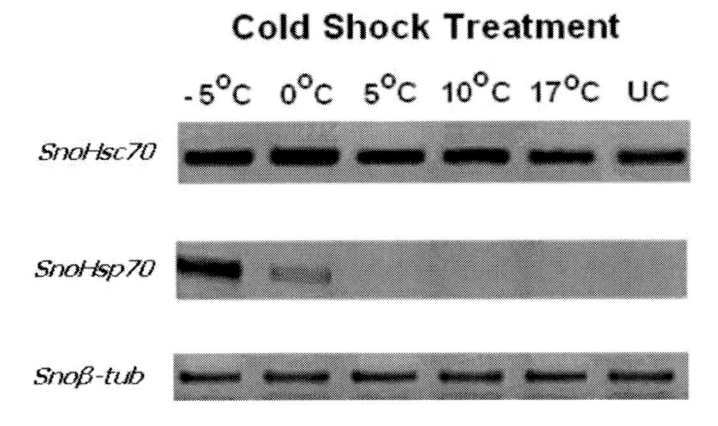

Figure 11. Semiquantitative RT-PCR expression analysis of *SnoHsc70/Hsp70* genes in nondiapausing larvae (5th instar-25 days since hatching) upon recovery 1 hour to LD 16:8 at 25 °C, after 1 hour cold shock at -5, 0, 5, 10 or 17°C. UC: untreated control. *Sno*β-tubulin was used as the control gene.

Hsp83 Gene

SnoHsp83 transcripts were amplified by semi-quantitative RT-PCR and examined using Real-Time PCR.Total RNA isolated from 25-day-old larvae that were subjected to various heat shock treatments was analyzed, using a pair of primers from the 3'-coding region of the gene. The expression of *SnoHsp83* was constitutive under unstressed conditions but was rapidly up-regulated under heat stress conditions. When 5th instar larvae were exposed to 40 °C, the transcripts were rapidly up-regulated (Figure 12. A, B). *SnoHsp83* is a highly heat-inducible gene and rapidly responding to stress.

There was a 2-fold increase in mRNA accumulation only after 15 min of exposure to heat stress and a 15-fold increase after 15 to 30 minutes of exposure, indicating that the induction of gene expression is manifested within few minutes after exposure to heat shock.

However, longer temperature exposure did not further up-regulate *SnoHsp83*, suggesting that *SnoHsp83* reached almost maximum levels of expression/accumulation within 30 min. Cold stress also affected the expression of *SnoHsp83* (Figure 13). In non-diapausing larvae, there was no induction of *SnoHsp83* gene expression when RNA was isolated from larvae immediately after cold stress. However, the induction was apparent only after a recovery period at ambient temperature (for 1 hour), when larvae were exposed at -5, 0, 5 °C for 1 hour. In non-diapausing larvae of *S. nonagrioides*, the *SnoHsp83* was up-regulated when the insects were exposed to an elevated temperature (40 ^{0}C).

The overall increase in the levels of *SnoHsp83* mRNA range from 2.0 to 15-fold on the basis of real-time PCR analysis. Also, exposure to -5, 0, or 5 ^{0}C caused *SnoHsp83* expression to elevated levels after recovery. In *S. nonagrioides* as previously reported for *SnoHsp20.8* and *SnoHsp70* transcripts (Gkouvitsas *et al.*, 2008; Gkouvitsas *et al.*, 2009) and for *Hsp90* in *Drosophila* and in *S. crassipalpis* (Yiangou *et al.*, 1997; Rinehart & Denlinger, 2000) the expression pattern detected by cold shock differed markedly from that induced by heat shock. Transcripts levels did not increase during the stress itself but only after a recovery period at 25 °C. This increase was evident 1 h after the shock and peaked 6 h after treatment.

In order to investigate whether environmental stress such as high temperature or cold may affect the *SnoHsp19.5* and *SnoHsp20.8* gene expression, non-diapausing larvae of *S. nonagrioides* were exposed to various temperatures. The *SnoHsp19.5* and *SnoHsp20.8* gene transcripts were amplified by semi-quantitative RT-PCR. The presence of *Sno20.8* mRNAs were also examined by a series of real-time PCR analyses. Transcript levels of *SnoHsp19.5* and *SnoHsp20.8* in non-diapausing larvae increased dramatically after the onset of heat shock at 40°C (Figure 14).

The *SnoHsp19.5* and *SnoHsp20.8* genes are highly heat-inducible genes and rapidly responding to stress. There was a 10-fold increase in mRNA accumulation in only after 15 min of exposure to heat stress, indicating that the induction of gene expression is manifested within few minutes after exposure to heat shock. A 10-fold increase was also apparent after 15 to 30 minutes of exposure at 40°C, showing that the expression activity remained at high levels.

Longer exposure to stress (40°C) had almost no additional effect, indicating that the mRNA accumulation of both genes reached almost maximum levels within 30 min.

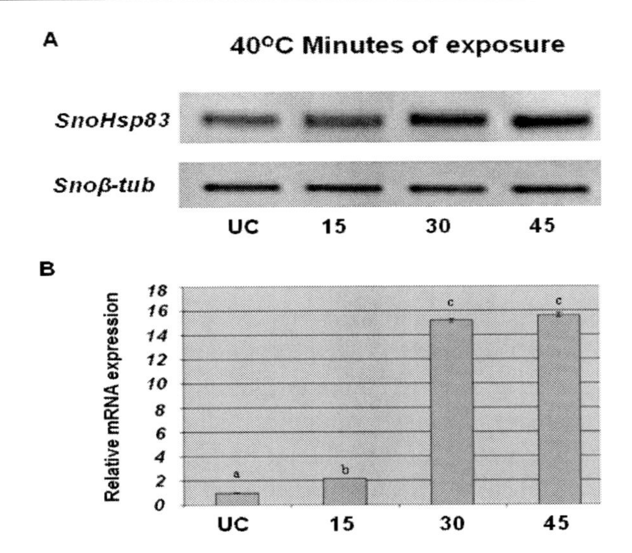

Figure 12. Expression of *SnoHsp83* in response to heat shock in non-diapausing larvae (5th instar-25 days since hatching), exposed to 40 °C for 15, 30, and 45 min. A. Semiquantitative RT-PCR expression analysis. *S. nonagrioides* β-tubulin was used as the control gene. B. Real-Time PCR analysis. UC: untreated control. The relative mRNA copy number of the *Snohsp83* gene transcripts is shown in relation to the number of *S. nonagrioides* β-tubulin transcripts for each cDNA sample. The error bars refer to the standard error of the mean.

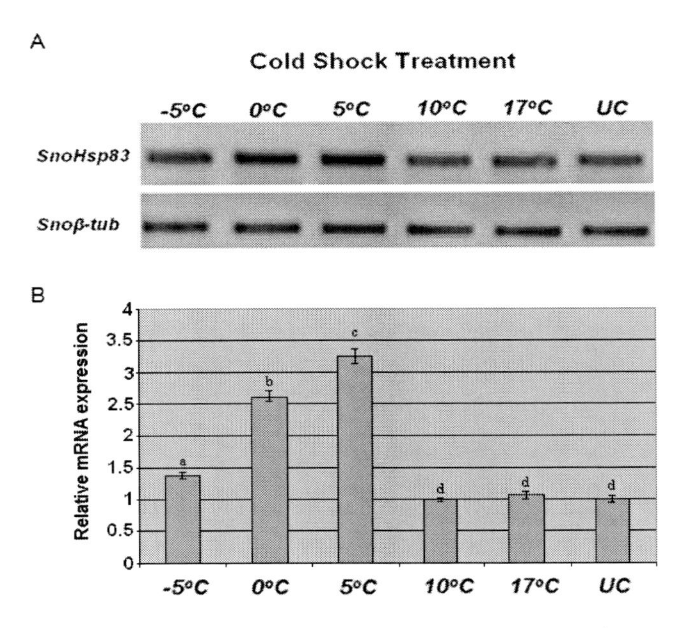

Figure 13. Expression analysis of *SnoHsp83* gene in non-diapausing larvae (5th instar-25 days since hatching) upon recovery 1 hour to LD 16:8 at 25 °C, after 1 hour cold shock at -5, 0, 5, 10 or 17°C. UC: untreated control. A. Semiquantitative RT-PCR expression analysis. *S. nonagrioides* β-tubulin was used as the control gene. B. Real-Time PCR analysis. The relative quantities indicate the levels of *Snohsp83* transcripts normalized to the internal standard *S. nonagrioides* β-tubulin. Each column is a mean±S.E.M. of six repeats. The differences in mRNA levels were assessed by ANOVA followed by Duncan's multiple comparison test ($P<0.05$) (columns flanked by different letters differ significantly).

Cold stress also affected the expression of *SnoHsp19.5* and *SnHspo20.8*. In non-diapausing larvae, there was no induction of gene expression when RNA was isolated from larvae immediately after cold stress. However the induction of both genes was apparent only after a recovery period at 25 °C (for 1 hour) under Long day conditions (16:8) , after 1 hour cold shock at -5, 0, 5°C or 10°C for *SnoHsp19.5* and 0, 5°C or 10°C for *SnoHsp20.8*. These results indicate that the two genes respond differently to different levels of heat stress when the larvae are cold-stressed for a short period of time and may open an interesting direction for future research.

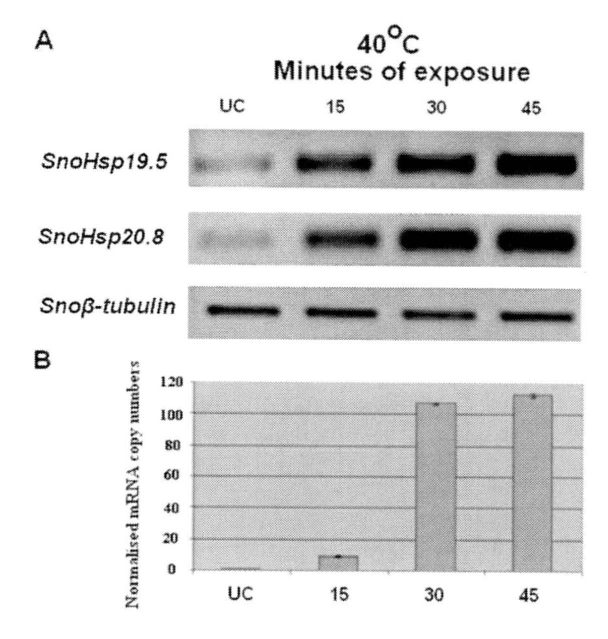

Figure 14. A. Semiquantitative RT-PCR expression analysis of *SnoHSP19.5* and *SnoHSP20.8* in response to heat shock in non-diapausing larvae (5th instar-25 days since hatching). Larvae exposed to 40 °C for 15, 30, and 45 min. UC: untreated control. *S. nonagrioides* β-tubulin was used as the control gene. B. Semiquantitative RT-PCR expression analysis of *Sno19.5* and *Sno20.8* genes in non-diapausing larvae (5th instar-25 days since hatching) upon recovery 1 hour to LD 16:8 at 25 °C, after 1 hour cold shock at -5, 0, 5, 10 or 17°C. UC: untreated control. *S. nonagrioides* β-tubulin was used as the control gene.

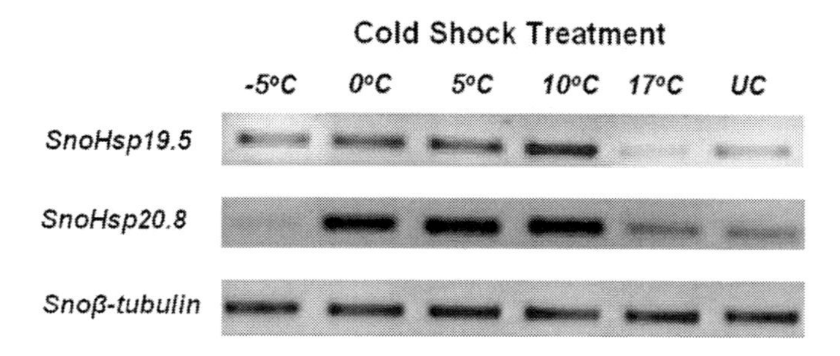

Figure 15. Semiquantitative RT-PCR expression analysis of *Sno19.5* and *Sno20.8* genes in non-diapausing larvae (5th instar-25 days since hatching) upon recovery 1 hour to LD 16:8 at 25 °C, after 1 hour cold shock at -5, 0, 5, 10 or 17°C. UC: untreated control. *S. nonagrioides* β-tubulin was used as the control gene.

Heat Shock Proteins and Their Role in Insect Developmental Processes

Diapause Programming and Heat Shock Protein Gene Expression

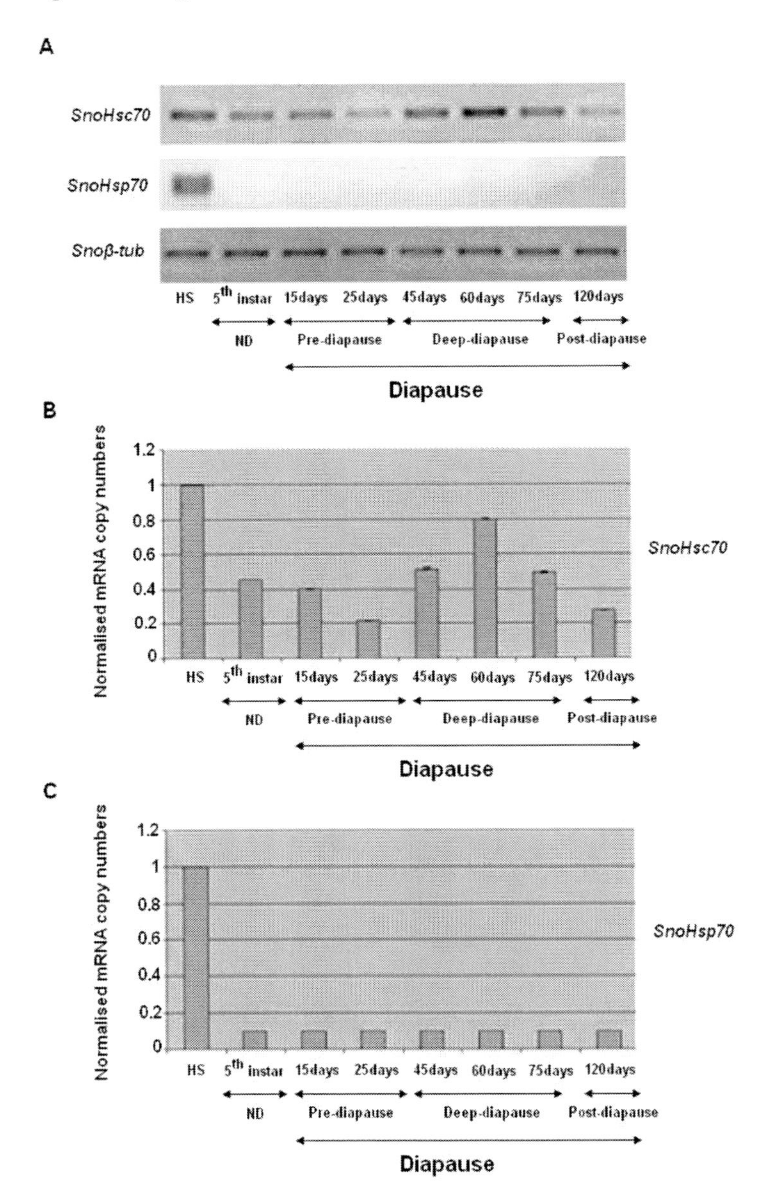

Figure 16. Expression of *SnoHsc70/Hsp70* in nondiapausing (5th instar-25 days since hatching) and diapausing larvae, reared under LD 16:8 at 25 °C and LD 10:14 at 25 °C, respectively. HS: non-diapausing larvae exposed to 40° C for 15 min. A. Semiquantitative RT-PCR expression analysis of *SnoHsc70/Hsp70* genes. *Sno*β-tubulin was used as the control gene. B. Real-Time PCR analysis of *SnoHsc70/Hsp70* genes. The relative mRNA copy number of the *SnoHsc70/Hsp70* genes transcripts are shown in relation to the number of *Sno*β-tubulin transcripts for each cDNA sample. The error bars refer to the standard error of the mean.

To investigate the transcriptional regulation of *SnoHsc/Hsp70* genes during diapause, RNA was isolated from 5th instar non-diapausing or diapausing larvae of *S. nonagrioides* and analyzed by semi-quantitative RT-PCR and Real-Time PCR assays. At 25°C, *SnoHsp70* transcripts were consistently low and there was almost no difference in the levels as a function of the developmental stage of non-diapause larvae or during diapause (Figure 16. A, C). In contrast, the *SnoHsc70* gene was upregulated during deep larval diapause (45-75 days since hatching). *SnoHsc70* showed constitutive expression in pre-diapause (15 days and 25 days since hatching) and post-diapause (120 days since hatching). At these stages, expression levels were similar to the levels observed in non-diapausing 5th instar larvae (Figure 16. A, B). Transcript accumulation increased in deep-diapause, showed a gradient increase with a peak in 60 day-old larvae and thereafter a gradual decrease. There was an apparent 2-fold increase in *Hsc70* expression levels when compared to S*noHsc70* mRNAs present in 5th instar larvae.

These results showed that the *SnoHsc70* transcript levels differed throughout the course of diapause. In contrast, the pattern of *SnoHsp70* transcript accumulation remained similar in all stages of diapause.

To determine whether the *SnoHsc*/Hsp70 genes respond to heat treatment during diapause, diapausing larvae (50 days since hatching) were transferred to 40° C for 1 h (Figure 17). We further tested the effects upon recovery to ambient temperature. The levels of *SnoHsp*70 transcripts were different between diapausing and non-diapausing larvae. Higher levels of *SnoHsp*70 mRNA accumulation were detected in non-diapausing larvae. Additionally, in non-diapausing larvae, high levels of *SnoHsp70* transcripts were immediately detected after heat shock. However, the decline in mRNA accumulation thereafter was rapid (1h). Further more, the transcripts gradually decreased over a period of 4h, reaching levels detected in non-diapausing larvae. In diapausing larvae, up-regulation of the gene was minimal; transcripts remained high levels 2 h after recovery but were undetected at 4h (Figure 17). In diapausing larvae, levels of *SnoHsc*70 transcripts were similar after recovery (0, 1, 2, 4h). In non-diapausing larvae the *SnoHsc*70 transcripts were high at 0 and 1h after heat treatment, but rapidly declined thereafter (Figure 17). These results show that the *SnoHsc70* gene responds to heat shock differently at the two stages tested. While it responds to heat shock in non-diapausing larvae, it does not respond in deep diapause, indicating that expression of the gene is also regulated by the diapause/non-diapause conditions.

Accumulated data from different species have resulted in contradictory conclusions. *Hsp70* is not upregulated as a function of diapause in *Drosophila triauraria* and *Lucilia sericata* (Goto *et al.*, 1998; Tachibana *et al.*, 2005) while it is slightly upregulated in the diapausing adults of Colorado potato beetle *Leptinotarsa decemblineata* (Yocum, 2001) and is only expressed in the diapausing pharate first instar larva of the gypsy moth *Lymantria dispar* after exposure to low temperature (Yocum *et al.*, 1991). In contrast, in a number of species including *S. crassipalpis* (Rinehart *et al.*, 2000), *L. decemlineata* (Yocum, 2001), *M. rotundata* (Yocum *et al.*, 2005), *D. antiqae* (Chen *et al.*, 2005), *C. suppressalis* (Sonoda *et al.,* 2006) and *B. mori* (Hwang *et al.*, 2005), the transcripts of *Hsp70* are up-regulated during diapause, even without thermal stress, and their expression persists throughout diapause. Rinehart *et al.* (2007) reported five additional species in which *Hsp70* was up-regulated during diapause (two dipterans and three lepidopteran species with diverse diapause programs: the obligate embryonic diapause of *Lymantria dispar*, the facultative larval diapause of *Ostrinia nubilalis*, and the facultative pupal diapause of *Manduca sexta*). *Hsp* up-regulation appears to be more common in pre-adult diapause than in adult (reproductive) diapause. The up-regulation is much more modest than seen in diapause occurring in pre-adult stages. It is also evident that the same *Hsps* are not consistently involved in diapause (Rinehart *et al.*, 2007). Our results show that *SnoHsc70* is induced during deep diapause. Only a few studies in insect diapause have reported on *Hsc70.* In *S. crassipalpis* the transcripts of *Hsc70* showed no difference in levels throughout the course of diapause and remained the same in diapausing and non-diapausing individuals (Rinehart *et al.*, 2000), and in *M. rotundata*, *Hsc70* is a normal part of both the diapause and post-diapause developmental programs (Yocum *et al.*, 2005).

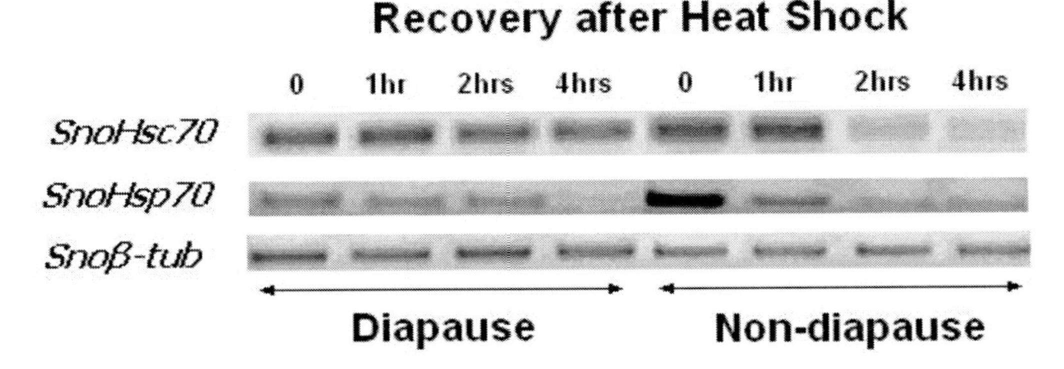

Figure 17. Semiquantitative RT-PCR expression analysis of *SnoHsc70/Hsp70* genes in nondiapausing (5th instar-25 days since hatching) and diapausing larvae (50 days since hatching) after heat shock at 40°C for 1 hour. Numbers indicate the hours upon recovery to LD 16:8 at 25 °C. Nondiapausing and diapausing larvae were reared under LD 16:8 at 25 °C and LD 10:14 at 25 °C, respectively. *Snoβ*-tubulin was used as the control gene.

Therefore different members of the *Hsp70* gene family respond differently during development or specific stages of insect diapause. Since most insects have a number of *Hsp70* members, the contradictory results observed in different species may be the consequence of different members being examined in the respective studies.

The regulation of the *SnoHsc70* is under strict developmental regulation. *SnoHsc70* may play a crucial role during the diapause of *S. nonagrioides*, most likely assisting the low but vital protein proper conformation during this specific developmental period. Possibly *SnoHsc70* transcripts during deep-diapause are hormonally controlled. Larval diapause is characterized by certain endocrine events, such as the presence of important juvenile hormone (JH) activity (Yin and Chippendale, 1979). According to Eizaguirre *et al.* (1998), JH is concerned in the maintenance of diapause in *S. nonagrioides.* In the presence of JH insect tissues maintain their developmental *status quo* when exposed to ecdysteroids (the molting hormones), whereas in the absence of JH, tissues change their commitment to that of the next metamorphic stage upon exposure to ecdysteroids (Riddiford, 1996; Nijhout, 1994). In *M. sexta*, the brain neuropeptide prothoracicotropic hormone (PTTH) stimulates a rapid increase in ecdysteroid hormone synthesis that is accompanied by general and specific increases in protein synthesis, including that of a 70 kDa cognate Heat shock protein (Hsc70) (Rybczynski and Gilbert, 2000). In this species, protein and mRNA data suggest that *Hsc*70 could be involved in a negative feedback loop regulating assembly of the ecdysone receptor complex.

*SnoHsc*70 could possibly be involved in a loop regulating the ecdysone receptor complex, as in *M. sexta.* The resolution of the function of *Hsc*70 in ecdysteroidogenesis or other cellular processes will require development of model systems in which *Hsc*70 levels can be readily manipulated under a variety of conditions.

SnoHsp83

SnoHsp83 showed constitutive expression in pre-diapause phase (15 days and 25 days since hatching) and post-diapause phase (90 days and 120 days since hatching). At these stages, the levels were similar to the ones observed at the 5th instar larvae of non-diapausing development (Figure 18. A,B). The transcript accumulation increased in deep-diapause period, showing a gradient increase with a peak at 60 day-old larvae and thereafter a gradual decrease. There was an apparent 2-fold increase in *Hsp83* expression levels during deep diapause when compared to *SnoHsp83* mRNAs present in 5th instar.

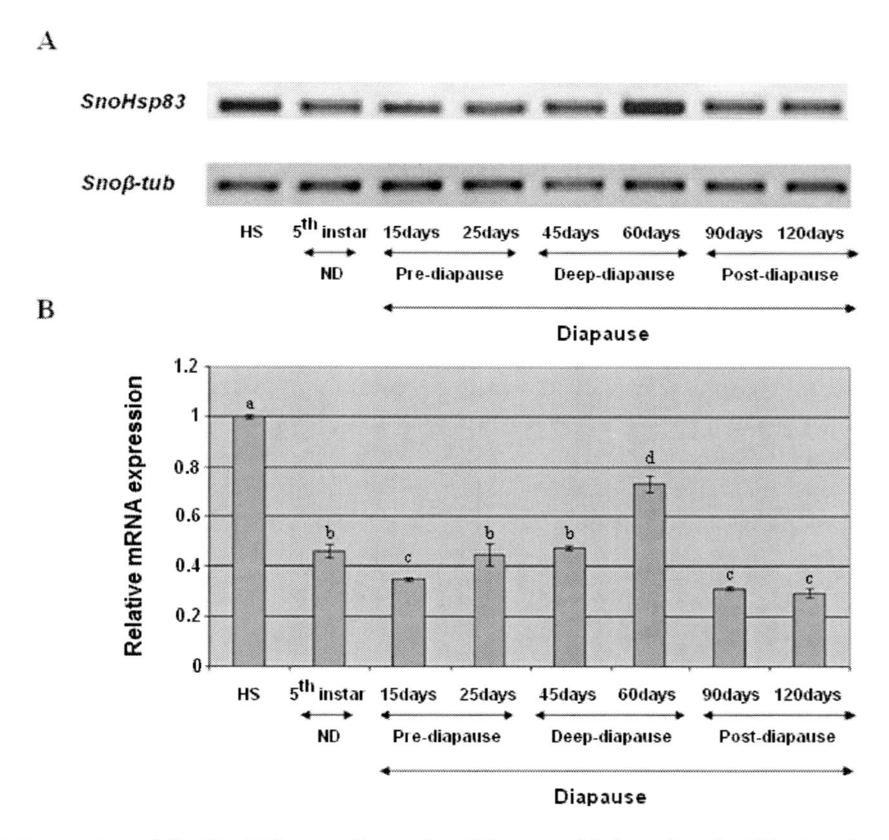

Figure18. Expression of *SnoHsp83* in non-diapausing (5th instar-25 days since hatching) and diapausing larvae, reared under LD 16:8 at 25 °C and LD 10:14 at 25 °C, respectively. A. Semiquantitative RT-PCR expression analysis of *SnoHsp83* gene. *S. nonagrioides* β-tubulin was used as the control gene. B. Real-Time PCR analysis of *SnoHsp83*gene. The relative mRNA copy number of the *SnoHsp83* gene transcripts is shown in relation to the number of *S. nonagrioides* β-tubulin transcripts for each cDNA sample. HS: non-diapausing larvae exposed to 40° C for 15 min.

To determine whether *SnoHsp83* respond to heat treatment through diapause, diapausing larvae (50 days since hatching) were transferred to 40° C for 1 h (Figure 19). We further tested the effects upon recovery to ambient temperature. The levels of *SnoHsp83* transcripts were different between diapausing and non-diapausing larvae. Higher levels of mRNA accumulation were evident in non-diapausing larvae. In non-diapausing larvae, high levels of *SnoHsp83* transcripts were apparent immediately after heat shock. However, the decline in mRNA accumulation thereafter was rapid (1h). The transcripts were gradually and slightly decreased over a period of 4h, reaching levels detected in non-diapausing larvae. In diapausing larvae, heat-shock increased the accumulation of mRNAs at lower level than that detected in diapausing larvae. The pattern of the gradual and slight decrease was also observed in non-diapausing larvae after recovery (Figure 19).

Our data indicate that *SnoHsp83* transcripts are regulated as a function of diapause, turning on as the larvae enters deep-diapause phase, and turning off in post- diapause phase. The regulation of *Hsp90* during diapause may involve ecdysteroids. Evidence exists for the up-regulation of *Hsp90* expression by ecdysteroids (Thomas and Lengyel, 1986). As a general rule, a decrease in the titers of both, hormones and ecdysteroids, is a signal for the metamorphic molt of holometabolous insects. Larval diapause is characterized by certain endocrine events, such as the presence of important juvenile hormone (JH) activity (Yin and Chippendale, 1979). In the presence of JH insect tissues maintain their developmental *status quo* when exposed to ecdysteroids (the molting hormones), whereas in the absence of JH, tissues change their commitment to that of the next metamorphic stage upon exposure to ecdysteroids (Riddiford, 1996; Nijhout, 1994). According to Eizaguirre *et al.* (1998), JH is concerned in the maintenance of diapause in *S. nonagrioides*. Corn stalk larvae developing under the short days (SD) conditions, undergo several extra larval molts which are associated with moderate JH titer and irregular rises of ecdysteroids; the titer rose before each extra larval molt (Eizaguire, *et al.,* 2005; Eizaguirre *et al.*, 2007). Although the mechanisms regulating *Hsp90* expression in such a manner remain unresolved, it is plausible that larval diapause characterized by the presense of ecdysteroids play a crucial role for the induction of its expression (Denlinger, 1985). In the *C. elegans*, dauer larvae are enriched in *Hsp90* gene transcripts and their products could interact with steroid hormone receptors (Dalley and Golomb, 1992; Cherkasova *et al.*, 2000). In *D. melanogaster* the presence of ecdysteroids leads to upregulation of *Hsp90* (Thomas and Lengyel, 1986) and this gene is critical for generating functional ecdysone receptors (Arbeitman and Hogness, 2000). These results consequently suggest that *Hsp90* is induced by ecdysteroid and is involved in developmental processes.

The up-regulation of *SnoHsp83*, as well as that of *SnoHsc70* may represent important molecular cues for corn borer diapause. It is very interesting that the pattern of *SnoHsp83* in diapause conditions increase in a similar fashion with the pattern of *SnoHsc70*. Both genes transcripts reached maximum level of accumulation in deep-diapausing larvae (60 days since hatching). Our results suggest that up-regulation of both *SnoHsc70* and *SnoHsp83* may are potent factors that have important roles in the mechanisms of diapause. *SnoHsc70* could possibly be involved in a loop regulating the ecdysone receptor complex (Gkouvitsas *et al.*, 2009). In *Drosoplila*, the steroid hormone 20-hydroxyecdysone coordinates the stages of development by activating a nuclear receptor heterodimer consisting of the ecdysone receptor, EcR, and the *Drosophila* RXR receptor, USP. The *Hsp90* and *Hsc70* are required *in vivo* for ecdysone receptor activity, and EcR is the primary target of the chaperone complex

(Arbeitman and Hogness, 2000). *In vivo* requirement for Hsp90 has been established for some steroid hormone receptors (Pratt and Toft, 1997). Monomeric glucocorticoid receptor and the closely related progesterone receptor are loaded onto Hsp90 by the Hsp70/ Hop-depended mechanism. A variation of the steroid receptor signaling pathway has been reported for the *Drosophila* ecdysone receptor, a member of the heterodimeric retinoid X receptor family. Hsp90 with its connection to the cellular signaling network may be particularly suited to such a function. Thus, the mechanisms of chaperone-mediated protein folding at a molecular level can be integrated with cellular processes and with the development of organisms (Young *et al.*, 2001).

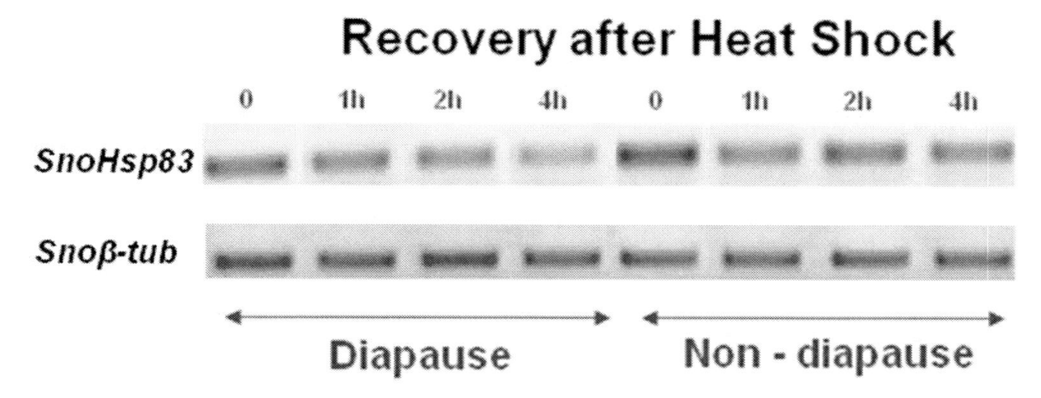

Figure 19. Semiquantitative RT-PCR expression analysis of *SnoHsp83* gene in non-diapausing (5th instar -25 days since hatcing) and diapausing larvae (50 days since hatching) after heat shock at 40°C for 1 hour. Numbers indicate the hours upon recovery to LD 16:8 at 25 °C. *S. nonagrioides* β-tubulin was used as the control gene.

It seems that like in *Drosophila*, *SnoHsp83* and *SnoHsc70* are possibly required for ecdysone activity. Consistent with this possibility is the up-regulation of the two genes in deep-diapause phase of *S. nonagrioides*, when extra larval molts go through. The up-regulation of *SnoHsp83*, as well as *SnoHsc70* may represent important molecular mechanisms for the survival of *S. nonagrioides* during deep-diapause. The regulation and function of the various Hsp families during diapause of corn stalk borer could provide insights into the molecular mechanisms involved in the diapause of this species.

SnoHsp19.5 and *SnoHsp20.8*

In order to investigate the transcriptional regulation of *SnoHsp19.5* and *SnoHsp20.8* genes during diapause, RNA was isolated from 5th instar non-diapausing or diapausing larvae and analyzed by semi-quantitative RT-PCR and Real-Time PCR assays. At 25°C, a non-stress temperature, the presence of mRNAs encoding sHsps is linked to the entire diapause period. As shown in Figure 20, the band intensity is no different from that of non-diapausing larvae. The expression of *SnoHsp19.5* gene begins with the onset of larval diapause, and the mRNAs are present until post-diapause (120 days since hatching). During the post-diapuse period, the larvae of *S. nonagrioides* will not initiate development. *SnoHsp20.8* transcript levels were increased in pre-diapause (15 days since hatching), down regulated during deep larval diapause and up-regulated in post-diapause stage. These results showed that *SnoHsp19.5* and *SnoHsp20.8* genes had different regulation regime throughout the course of diapause.

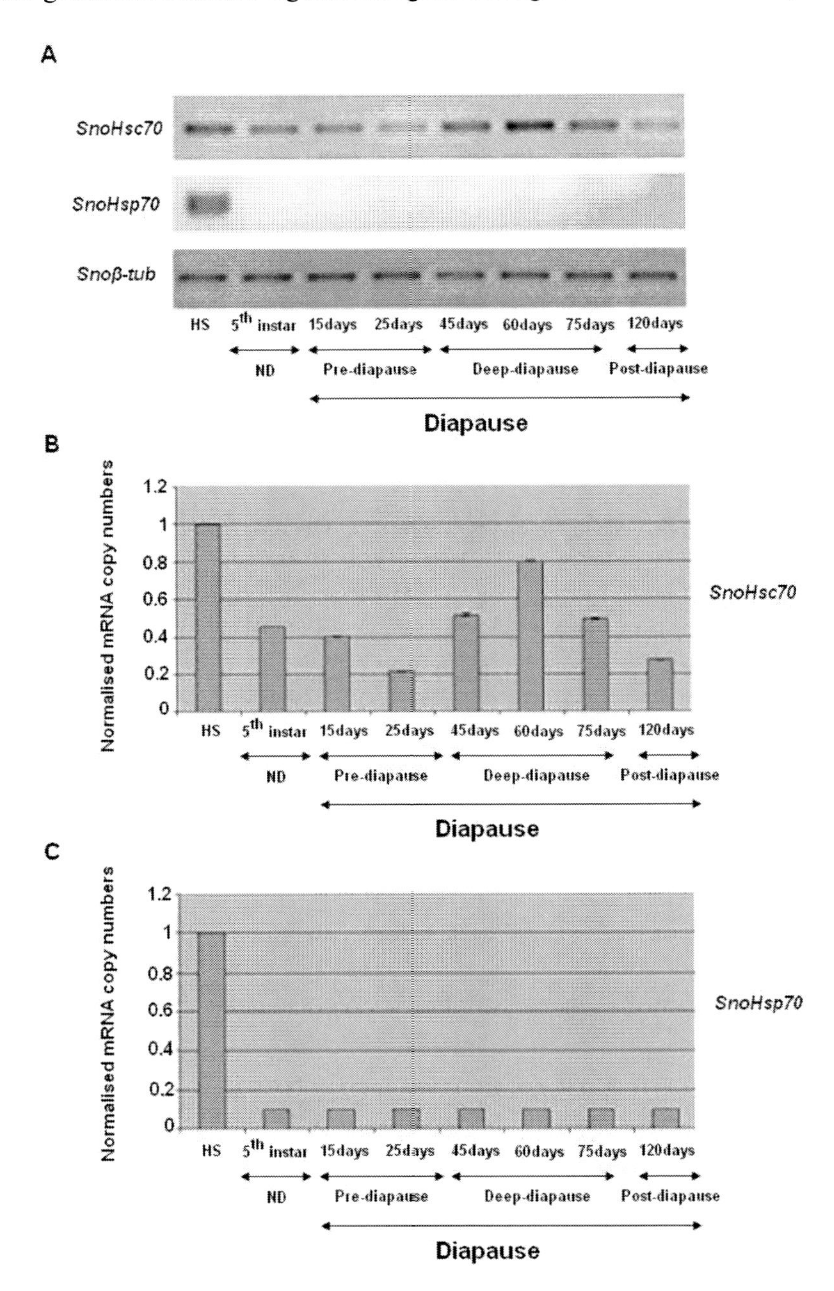

Figure 20. Expression of *Sno19.5* and *Sno20.8* in non-diapausing (5o instar-25 days since hatching) and diapausing larvae, reared under LD 16:8 at 25 °C and LD 10:14 at 25 °C, respectively. *S. nonagrioides* β-tubulin was used as the control gene. HS: non-diapausing larvae exposed to 40° C for 15 min. A. Semiquantitative RT-PCR expression analysis of *Sno19.5* and *Sno20.8* genes. B. Real-Time PCR analysis of *Sno20.8* gene. The relative mRNA copy number of the *Sno20.8* gene transcripts is shown in relation to the number of *S. nonagrioides* β-tubulin transcripts for each cDNA sample. The error bars refer to the standard error of the mean.

In order to define detailed transcriptional profile of the two genes during diapause or non-diapause conditions, we induced the expression for 1h at 40° C (Figure 21). Both genes were highly and rapidly expressed at about similar levels. However, under non-diapause conditions, *SnoHsp20.8* transcripts remained for longer period than *SnoHsp19.5* mRNAs. This phenomenon was reversed when larvae were in diapause. The *SnoHsp19.5* transcripts remained for longer period than *SnoHsp20.8* mRNAs, indicating that the former should play a crucial role in diapause.

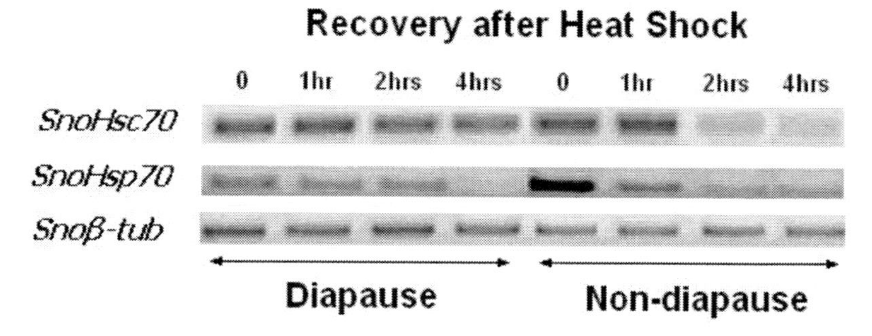

Figure 21. Semiquantitative RT-PCR expression analysis of *Sno19.5* and *Sno20.8* genes in non-diapausing (5o instar) and diapausing larvae (50 days after hatching) upon recovery to LD 16:8 at 25 °C, after a heat shock at 40°C for 1 hour. Numbers indicate the hours after returning. *S. nonagrioides* β-tubulin was used as the control gene.

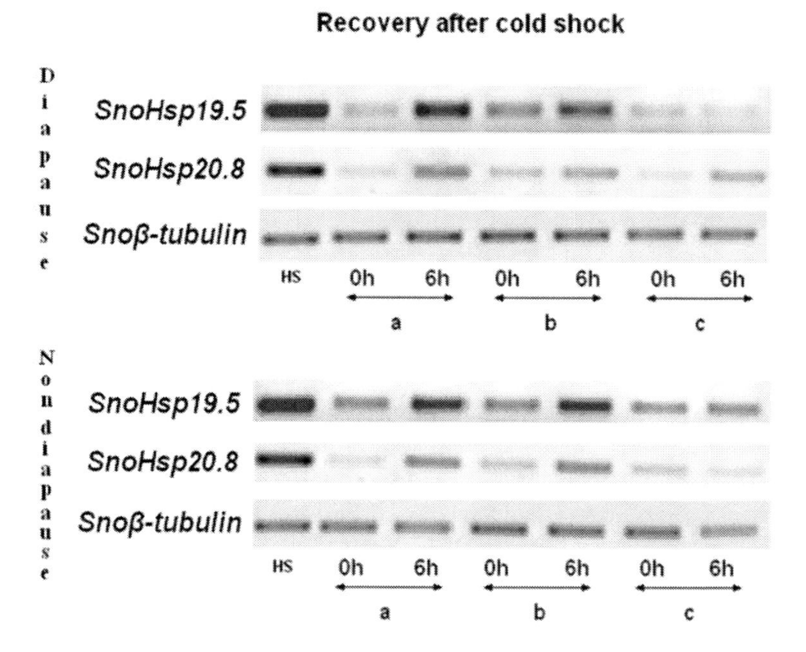

Figure 22. Semiquantitative RT-PCR expression analysis of *Sno19.5* and *Sno20.8* genes after recovery from cold shock in diapausing larvae (50 days after hatching) and nondiapausing (25 days since hatching) upon recovery after cold shock at -5°C for: a =15 min, b = for 30 min and c= 60 min. Numbers indicate the hours after returning to LD 16:8 at 25 °C. HS = non-diapausing larvae exposed to 40 °C for 15 min. *S. nonagrioides* β-tubulin was used as the control gene. A. Diapausing larvae and B. Non- diapausing larvae.

We additional tested the effects of cold shock (-5°C for 15, 30 or 60 min) on the expression pattern of *SnoHsp19.5* and *SnoHsp20.8* genes in diapausing and non-diapausing larvae (Figure 22) and we observed higher levels of transcripts after 6 hours recovery. There was an obvious induction/mRNA accumulation within 60 min after recovery at 25°C when diapausing or non diapausing insects were cold-stressed for 15 or 30 min. However, when insects were stressed for longer period (60 min) at 5°C there was no obvious induction/mRNA accumulation after recovery except for *SnoHsp20.8* gene in diapausing larvae (Figure 22).

Our results show clearly that the levels of expression of *SnoHsp19.5* and *SnoHsp20.8* regulated in different way in response to diapause. In diapausing larvae of *S. nonagrioides*, *SnoHsp19.5* transcripts were consistently expressed throughout diapause development while *SnoHsp20.8* transcripts undergo regulated changes. *SnoHsp20.8* transcripts are relatively high during the initiation and termination phases of diapause. This pattern of accumulation clearly defines the early stages of diapause as well as the culmination of the diapause process.

SnoHsp19.5 and *SnoHsp20.8* may have distinctive roles in the initiation, continuation and termination of the dynamic process of diapause. Since *SnoHsp19.5* transcripts remained at high levels after heat treatment in the diapause, it is plausible to consider that this gene could play a role in the regulation of diapause. In contrast, *SnoHsp20.8* transcripts remained at low levels after heat treatment in the diapause while remained at high levels after heat treatment in non-diapausing larvae. This clear distinction in gene expression and accumulation of the two genes in diapause or non-diapause larvae vividly indicates the distinctive roles of the two genes during the process of diapause and development. Members of the sHsps families are up-regulated during dormancies in several non-insect organisms (*Saccharomyces cerevisiae*, *Blastocladiella emersonii*, *Artemia franciscana*, *Quercus suber*, *Caenorhabditis elegans*). Denlinger et al. (2001) proposed two hypotheses to explanation for their response: (1) small *Hsp*s are involved directly in the cell cycle arrest during dormancy; (2) they function as part of the increased stress resistance of dormant individuals.

The former hypothesis appears to be invalid for the larval diapause of *S. nonagrioides* because expression of the *Hsp20.8* transcripts is not up-regulated as a function of diapause. Neither is this the case in *D. triauraria* (Goto *et al.*, 1998; Goto and Kimura, 2004) nor in the gypsy moth *Lymantria dispar* (Yocum *et al.*, 1991; Denlinger *et al.*, 1992). Thus, up-regulation of small Hsps would not be a common feature of diapause insects. However, it is still necessary to investigate the expression of other copies of other members of small *Hsp*s before validating this conclusion. On the other hand, in *D. triauraria*, *hsp23* and *hsp26* are undetectable irrespective of the diapause state at normal temperatures (Goto *et al.*, 1998; Goto and Kimura 2004). The levels of *hsp23* were also consistently low irrespective of the diapause status in *L. sericata* (Tachibana *et al.*, 2005). A possible role for small heat shock proteins in diapause is the involvement in the regulation of the cell cycle arrest (Tammarielo and Denlinger, 1998). Denlinger *et al.* (2001) suggested that small HSPs are involved in protecting organisms against low temperature stress and in the regulation of cell cycle arrest. Members of the small Hsps families are upregulated during dormancies in several non insect organisms (Bonato *et al.*, 1987; Pla *et al.*, 1998; Cherkasova et al., 2000). The fact that sHsps are up-regulated during diapause in a variety of insect species representing different diapause stages suggests that sHsps are key players in the overwintering response of many insects.

In *S. crassipalpis*, a sHsp (Hsp23) is highly up regulated during fly diapause and this occurs simply because the pupa enters diapause and is not a function of temperature stress

(Rinehart *et al.*, 2007). Rinehart *et al.* (2007) propose that up-regulation of sHsps during diapause is a major factor contributing to cold-hardiness of overwintering insects. They consider that up-regulation of sHsps provide a "back-up" mechanism guaranteeing that development will remain arrested during diapause. By contrast, *sHsp* genes were not up-regulated in response to diapause in *D. triauraria, L. sericata* and *C. suppressalis* (Goto and Kimura, 2004; Tachibana *et al.* 2005; Sonoda *et al.*, 2006). Our results support both data. Other *sHsps* are induced while others are suppressed. In the question why are there so many *sHsps*, in *S. nonagrioides* seems that each of *SnoHsp19.5* and *SnoHsp19.5* have a specific function. Most likely, the different *sHsps* are interacting differently with specific groups of proteins (Sun and MacRae, 2005). Our results suggest that these two *sHsps* contribute to diapause and to cold tolerance in slightly different ways. The expression of *SnoHsp19.5* gene was consistently expressed in response to diapause and up-regulated in stress conditions, while *SnoHsp20.8* gene was down regulated in deep diapause and up-regulated at the termination of diapause. Thus, up-regulation of small heat shock proteins would not be a common feature of diapause insects.

Conclusion

The Hsps genes play important roles to several physiological processes of insects. In *S. nonagrioides* the *Hsp*s genes are implicated not only to abiotic stresses response but also to developmental processes, like diapause programming.The differential expression of *SnoHsc*70 and *SnoHsp*70 during diapause and in response to high and low temperatures indicates that individual members of the 70 kDa heat shock protein family may well have varying roles in diapause of *S. nonagrioides.* The upregulation of *SnoHsp83*, as well as that of *SnoHsc70* could represent important molecular cues for corn borer diapause. It is very interesting that the two genes display a similar pattern when larvae are under diapause conditions. The *SnoHsp19.5* and *SnoHsp20.8* contribute to diapause and to cold response in slightly different ways. The *SnoHsp19.5* gene was consistently expressed in response to diapause, while *SnoHsp20.8* gene was down-regulated in deep diapause and up-regulated at the termination of diapause. Therefore, an analyses of sHsps can provide significant information for understanding diapause in insects. The results presented here for the corn stalk borer and other examples from the literature clearly demonstrate that the regulation of the heat shock response is not a simple on/off reaction but is finely tuned to developmental and environmental conditions. The regulation and function of the various Hsp families during diapause of corn stalk borer could provide insights into the molecular mechanisms involved in the diapause of this species.

Acknowledgments

Reviewed by: Dimitra Milioni, Assistant Professor, Agricultural Biotechnology Department, Agricultural University of Athens, Iera Odos 75, 11855, Athens, Greec, Tel. +30-2105294348, dmilioni@aua.gr.

References

Altaratz, M., Applebaum, Sh.W., Richard, D.S., Gilbert, L.I., Segal, D. (1991). Regulation of juvenile hormone synthesis in wild-type and apterous mutant *Drosophila. Mol. Cell. Endocrinol.* 81, 205-216.).

Altschul, S.F., Madden, T.L., Schaffer, A.A., Xhang, J., Zhang, Z, Miller, W.and Lipman, D.J. (1997) Gapped BLAST and PSI-BLAST: a new generation of protein database search programs. *Nucleic Acids Res* 25: 3389-3402.

Arbona, M., de Frutos, R. and Tanguay, R.M. (1993) Transcriptional and translational study of the *Drosophila subobscura hsp83* gene in normal and heat-shock conditions. *Genome* 36, 694–700.

Atungulu, E., Tanaka, H., Fujita, K., Yamamoto, K., Sakata, M., Sato, E., Hara, M., Yamashita', T., Suzuki, K.(2006). A Double Chaperone Function of the sHsp Genes against Heat-Based Environmental Adversity in the Soil-Dwelling Leaf Beetles. *Journal of Insect Biotechnology and Sericology* 75, 15-22.

Benedict, M.Q., Levine, B.J., Ke, Z.X., Cockburn, A.F. and Seawright, J.A. (1996) Precise limitation of concerted evolution to ORFs in mosquito *Hsp82* genes. *Insect Mol Biol* 5, 73–79.

Blackman, R.K. and Meselson, M. (1986) Interspecific nucleotide sequence comparisons used to identify regulatory and structural features of the *Drosophila hsp82* gene. *J Mol Biol* 188, 499–515.

Bonato, M.C., Silva, A.M., Gomes, S.L., Maia, J.C., Juliani, M.H. (1987). Differential expression of heat-shock proteins and spontaneous synthesis of HSP70 during the life cycle of *Blastocladiella emersonii. European Journal Biochemistry* 163, 211–220.

Boorstein, W.R., Ziegelhoffer, T. And Craig, E.A. (1994) Molecular evolution of the Hsp70 multigene family. *J. Mol. Evol.* 38, 1–17.

Buchner, J. (1999) Hsp90 – a holding for folding. *Trends Biochem Sci* 24,136–141.

Caplan, A.J. (1999) HSP90's secrets unfold: new insights from structural and functional studies. *Trends Cell Biol* 9, 262-268.

Cavener, D.R. (1987) Comparison of the consensus sequence flanking translational start sites in *Drosophila* and vertebrates. *Nucleic Acids Res* 15, 1353–1361.

Chen, B., Kayukawa, T., Monteiro, A., Ishikawa, Y. (2005). The expression of the *HSP90* gene in response to winter and summer diapauses and thermal-stress in the onion maggot, *Delia antique. Insect Molecular Biology* 14, 697-702.

Chen, B., Kayukawa, T., Monteiro, A. and Ishikawa, Y. (2006) Cloning and characterization of the *Hsp70* gene, and its expression in response to diapauses and thermal stress in the onion maggot, *Delia antiqua. J. Bioch.. Mol. Biol.* 39, 749-758.

Cherkasova, V., Ayyadevara, S., Egilmez, N., Shmookler, Reis, R. (2000). Diverse *Caenorhabditis elegans* genes that are upregulated in dauer larvae also show elevated transcript levels in long-lived, aged, or starved adults. *Journal of Molecular Biology* 300, 433–448.

Craig, E.A., Inglolia, T.D. and Manseau, L.J. (1983) Expression of *Drosophila* heat-shock cognate genes during heat shock and development. *Dev. Biol.* 99, 418-426.

Craig, E. A., Gambill, B.D. and Nelson R. J. (1993) Heat shock proteins: molecular chaperones of protein biogenesis. *Microbiol Mol Biol Rev* 57, 402-414.

Dalley, B.K. and Golomb, M. (1992) Gene expression in the *Caenorhabditis elegans* dauer larva: developmental regulation of hsp90 and other genes. *Dev Biol* 151, 80- 90.

Danks HV. (1987). Insect Dormancy: An ecological Perspective (Biological Survey, Ottava, Canada).

Denlinger, D. L., (1985). Hormonal control of diapause. In Comprehensive insect physiology, biochemistry and pharmacology, eds. Kerkut G. A. and Gilbert L. I., Vol.8, pp 353–412. Pergamon, Oxford.

Denlinger, D.L., Lee, R.E., Yocum GD, Kukal, O. (1992). Role of chilling in the accuisition of cold tolerance and the capacitation to express stress proteins in diapausing pharate larvae of the gypsy moth *Lymantria dispar*. *Archives of Insect Biochemistry and Physiology* 21, 271-280.

Denlinger, D.L. (2002). Regulation of diapause. *Annual Review of Entomology* 47, 93-122.

Denlinger, D.L., Rinehart JP, Yocum GD. (2001). Stress proteins: a role in insect diapause? In: *Insect Timing: Circadian Rhythmicity to Seasonality*. Ed. by Denlinger, D.L., Giebultowicz, J., Saunders, D.S., Elsevier, Amsterdam, 155–171.

Denlinger, D.L., Yocum, G.D. and Rinehar, J.P. (2005) Hormonal control of diapause. In: Comprehensive Molecular Insect Science, vol. 3. Gilbert, L.I., Iatrou, K.,and Gill, S. (Eds). Elsevier, Amsterdam, pp.615-650.

Ehrnsperger, M., Graber, S., Gaestel, M., Buchner, J. (1997). Binding of non-native protein to Hsp25 during heat shock creates a reservoir of folding intermediates for reactivation, *EMBO Journal* 16, 221–229.

Eizaguirre, M., Lopez, C., Asin, L., Albajes, R. (1994). Thermoperiodism, pfotoperiodism and sensitive stage in the diapause induction of *Sesamia nonagrioides* (Lepidoptera: Noctuidae). *Journal of Insect Physiology* 40, 113-119.

Eizaguirre, M., Lopez, C., Asin, L. and Albajes, R. (1994) Thermoperiodism, pfotoperiodism and sensitive stage in the diapause induction of *Sesamia nonagrioides* (Lepidoptera: Noctuidae). *J. Insect. Physiol.* 40, 113-119.

Fantinou, A.A., Karandinos, M.G., Tsitsipis, J.A. (1995). Diapause induction in the *Sesamia nonagrioides* (Lepidoptera: Noctuidae) effect of photoperiod and temperature. *Environmental Entomology* 24, 1458-1466.

Fantinou, A.A., Tsitsipis, J.A. and Karandinos, M.G. (1996) Effects of short- and long-day photoperiods on growth and development of *Sesamia nonagrioides* (Lepidoptera: Noctuidae). *Environm. Entomol.* 25, 1337-1343.

Fantinou, Tsitsipis, J.A., A.A., Karandinos, M.G. (1998). Diapause termination in *Sesamia nonagrioides* (Lepidoptera: Noctuidae) under laboratory and field conditions. *Environmental Entomology* 27, 53-58.

Falquet, L., Pagni, M., Bucher, P., Hulo, N., Sigrist, C.J., Hofmann, K., Bairoch, A. (2002). The PROSITE database, its status in 2002. *Nucleic Acids Research* 30, 235–238.

Feder, J.H., Rossi, J.M., Solomon, N., Linquist, S. (1992). The consequences of expressing hsp70 in *Drosophila* cells at normal temperatures. *Genes Development* 6, 1402-1413.

Feder, M. E., Hofmann, G. E. (1999). Heat-shock proteins, molecular chaperones, and the stress response: evolutionary and ecological physiology. *Annu. Rev. Physiol.* 61, 243-282.

Fink, A.L. (1999). Chaperone-mediated protein folding. *Physiol Rev.* 79, 425-49.

Felsenstein, J. (2001). PHYLIP (Phylogeny Inference Package) version 3.6alpha. Distributed by the author. Department of Genetics, University of Washington, Seattle.

Frohman, M.A. (1990). RACE: Rapid amplification of cDNA ends. In PCR Protocols: A Guide to Methods and Applications. Eds. Innis, M.A., Gelfand, D.H., Sninsky, J.J., and White, T.J.. Academic Press, London, pp 28-38.

Flanagan, R.D., Tammariello, S.P., Joplin, K.H., Cirka-Ireland, R.A., Yocum, G.D., Denlinger, G.D. (1998) Diapause specific expression in pupae of the flesh fly *Sarcophaga crassipalpis*. *Proc Natl Acad Sci USA* 95, 5616-5620.

Hartl, F.U., Hayer-Hartl, M. (2002). Molecular chaperons in the cytosol: from nanscent chain to folded protein. *Science* 295, 1852-1266.

Hayward, S.A.L., Rinehart, J.P., Denlinger, D.L. (2004). Desiccation and rehydration elicit distinct heat shock protein transcript responses in flesh fly pupae. *Journal of Experimental biology* 207, 963-971.

Hilal, A. (1977). Mise en evidence d' un etat diapause vrae chez *Sesamia nonagrioides* Lef. (Lepidoptera-Noctuidae). *C. R. Acad. Sci. Ser. D. Paris* 285, 365-367.

Galichet, P.F. (1982) Hibernation d'une population de *Sesamia nonagrioides* Lef. (Lep. : Noctuidae) en France Meridionale. Agronomie (Paris). 2, 561-566.

Garnier, J., Gibrat, J.F., Robson, B. (1996). GOR secondary structure prediction method version IV. In Methods in Enzymology. Ed. R.F. Doolittle, vol 266, 540-533.

Gething M.J. ed. (1997). Guidebook to Molecular Chaperones and protein-Folding Catalysts. Oxford, UK: Oxford Univ. Press.

Gkouvitsas, T., Kontogiannatos, D. and Kourti, A. (2008). Differential expression of two small Hsps during diapause in the corn stalk borer *Sesamia nonagrioides* (Lef.). *J Insect Physiol* 54, 1503-1510.

Gkouvitsas, T., Kontogiannatos D. and Kourti, A. (2009). Cognate Hsp70 gene is induced during deep diapause in the moth *Sesamia nonagrioides*. *Insect Mol Biol* 14, 697-702.

Gkouvitsas, T., Kontogiannatos, D., Kourti, A. (2009). Expression of the *Hsp83* gene in response to diapause and thermal stress in the moth *Sesamia nonagrioides*. *Insect Mol. Biol.* 18, 759-768.

Goto, S.G., Yoshida, K.M., Kimura, M.T. (1998). Accumulation of Hsp70 mRNA under environmental stresses in diapausing and nondiapausing adults of *Drosophila triauraria*. *Journal of Insect Physiology* 44, 1009–1015.

Goto, S.G., Kimura, M.T. (2004). Heat-shock-responsive genes are not involved in the adult diapause of *Drosophila triauraria*. *Gene* 326, 117–122.

Gupta, R.S. and Golding, G.B. (1993). Evolution of Hsp70 gene and its implications regarding relationships between archaebacteria, eubacteria, and eukaryotes. *J. Mol. Evol.* 37, 573-582.

Gupta, R.S. (1995). Phylogenetic analysis of the 90 kD heat shock family of protein sequences and an examination of the relationship among animals, plants, and fungi species. *Mol Biol Evol* 12, 1063–1073.

Hartl, F.U. (1996). Molecular chaperones in cellular protein folding. *Nature* 381, 571–580.

Hartl, F.U. and Hayer-Hartl, M. (2002) Molecular chaperons in the cytosol: from nascent chain to folded protein. *Science* 295: 1852-1266.

Hendrick, J. P. and Hartl, F. (1993) Molecular Chaperone Functions of Heat-Shock Proteins. *An Rev Bioch* 62, 349-384.

Hung, J.J., Cheng, T.J., Chang, M.D., Chen, K.D., Huang, H.L. and Lai, Y.K. (1998). Involvement of heat shock elements and basal transcription elements in the differential induction of the 70-kDa heat shock protein and its cognate by cadmium chloride in 9L rat brain tumor cells. *J Cell Biochem*. 71, 21-35.

Hwang, J.S., Go, H.J., Goo, T.W., Yun,E.Y., Choi, K.H., Seong, S.I., Lee, S.M., Lee, B.H., Kim, I., Taehoon Chun, T. and Kang, S.W. (2005). The analysis of differentially expressed novel transcripts in diapausing and diapause-activated eggs of *Bombyx mori. Arch. Insect Biochem. Physiol.* 59, 197-201.

Ingolia, T.D., Craig, F.A. (1982). Four small *Drosophila* heat shock proteins are related to each other and to mammalian α-crystallin. *Proceedings of National Academic Science USA* 79, 2360-2364.

Ireland, R.C., Berger, E.M. (1982). Synthesis of the low molecular weight heat shock proteins stimulated by 20-hydroxyecdysone in a cultured *Drosophila* cell line. *Proceedings of National Academic Science USA* 79, 855–859.

James, M., Crabbe, M.J.C., Goode, D. (1994). α-Crystallin: chaperoning and aggregation. *Biochemistry Journal* 297, 653-654.

Joplin, K.H., Denlinger, D.L. (1990). Developmental and tissue specific control of the heat shock induced 70 kDa related proteins in the flesh fly *Sarcophaga crassipalpis. Journal of Insect Physiology* 36, 239.

Karouna-Renier, N.K., Yang, W.J. and Ranga, R.K.(2003) Cloning and characterization of a 70 kDa heat shock cognate gene (*HSC70*) from two species of *Chironomus*. *Insect. Mol. Biol* .12, 19–26.

Kiang, J.G., Tsokos, G.C. (1998). Heat shock protein 70 kDa: molecular biology, biochemistry, and physiology. *Pharmacol. Ther.* 80, 183–201.

Kostal, V. (2006). Eco-physiological phases of insect diapause. *Journal of Insect Physiology* 52, 113-127.

Kourti, A. (2006). Mitochondrial DNA Restriction Map and Cytochrome c Oxidase Subunits I and II Sequence Divergence of Corn Stalk Borer *Sesamia nonagrioides* (Lepidoptera: Noctuidae). *Biochem Genet* 44, 321-32.

Kregel, K.C. (2002). Heat shock proteins: modifying factors in physiological stress responses and acquired thermotolerance. *J. Appl. Physiol.* 92, 2177–2186.

Kurtz, S., Rossi, J., Petko, L., Lindquist, S. (1986). An ancient developmental induction heat-shock proteins induced in sporulation and oogenesis. *Science* 231, 1154–1157.

Landais, I., Pommet, J., Mita, K., Nohata, J., Gimenez, S., Fournier, P., Devauchelle, G., Duonor-Cerutti, M. and Ogliastro, M. (2001). Characterization of the cDNA encoding the 90 kDa heat-shock protein in the Lepidoptera, *Bombyx mori* and *Spodoptera frugiperda*. *Gene* 271, 223–231.

Lee, S.M., Lee, S.B., Park, C.H., Choi, J. (*2006). Expression* of *heat shock protein* and hemoglobin genes in *Chironomus tentans* (Diptera: Chironomidae) larvae. *Chemosphere* 65, 1074-81.

Linquist, S. (1986). The heat shock response. *Ann. Review Biochem*. 55, 1151–1191.

MacRae, T.H., 2000. Structure and function of small heat shock/α-crystallin proteins: established concepts and emerging ideas. Cell Molecular Life Science 57, 899–913.

Mahroof, R., Zhu, K.Y., Neven, L., Subramanyam, B. and Bai, J. (2005). Expression patterns of three heat shock protein 70 genes among developmental stages of the red flour beetle, *Tribolium castaneum* (Coleoptera: Tenebrionidae). *Comp. Biochem. Physiol.* A 141, 247–256.

Mason, P.J., Hall, L., Gauz, J. (1984). The expression of heat shock genes during normal development in *Drosophila melanogster. Mol. Gen. Genet.* 194, 73-78.

Meacham, G.C., Patterson, C., Zhang, W., Younger, J.M., and Cyr, D.M. (2001). The Hsc70 co-chaperone CHIP targets immature CFTR for proteasomal degradation. *Nat Cell Biol* 3, 100-105.

Narberhaus, F. (2002). Alpha-crystallin-type heat shock proteins: socializing minichaperones in the contex of a multichaperone network. *Microbiology and Molecular Biology Reviews* 66, 64-93.

Nathan, D.F., M.H. Vos, and S. Lindquist (1997). In vivo functions of the *Saccharomyces cerevisiae* Hsp90 chaperone. *Proc Natl Acad Sci USA.* 94, 12949–12956.

Nijhout, H.F. (1994).Genes on the wing. *Science* 265: 44-45.

Nover, L. And Scharf, K.D. (1997). Heat stress proteins and transcription factors. *Cell Mol. Life Sci.* 53: 80-103.

Pla, M., Huguet, G., Verdaguer, D., Puigderrajols, P., Llompart, B., Nadal, A., Molinas, M. (1998). Stress proteins co-expressed in suberized and lignified cells and in apical meristems. *Plant Science* 139, 49–57.

Parsell, D.A. and Lindquist, S. (1993). The function of heat-shock proteins in stress tolerance – degradation and reactivation of damaged proteins. *Annu Rev. Genet.* 27, 437–496.

Picard, D., Khursheed B., Garabedian, M.J., Fortin, M.G., Lindquist, S. and Yamamoto, K.R. (1990). Reduced levels of hsp90 compromise steroid receptor action *in vivo. Nature* 348, 166 – 168.

Picard, D. (2002). Heat-shock protein 90, a chaperone for folding and regulation. *Cell Mol Life Sci* 59, 1640–1648.

Pratt, W.B. and Toft, D.O. (2003). Regulation of signaling protein function and trafficking by the hsp90/hsp70-based chaperone machinery. *Exp Biol Med (Maywood)* 228, 111–133.

Qin, W., Tyshenko, M.G., wu, B.S., Walker, V.K., Robertson, R.M. (2003). Cloning and characterization of a member of the hsp70 gene family from *Locusta migratoria*, a highly thermotolerant insect. *Cell Stress Chaperones* 8, 144-152.

Ramsey, A.J., Russell, L.C, Whitt, S.R. and Chinkers, M. (2000). Over-lapping sites of tetratricopeptide repeat protein binding and chaperone activity in heat shock protein 90. *J Biol Chem* 275, 17857-17862.

Rinehart, J.P. and Denlinger, D.L. (2000). Heat -shock protein 90 is down-regulated during pupal diapause in flesh fly, *Sarcophaga crassipalpis,* but remains responsible to thermal stress. *Insect Mol Biol 9, 641-645.*

Rinehart, J.P, Yocum, G.D., Denlinger, D.L. (2000). Developmental upregulation of inducible hsp70 transcript, but not the cognate form, during pupal diapause in the flesh fly, *Sarcophaga crassipalpis. Insect Biochem Mol Biol* 30, 515–521.

Rinehart, J.P., Li, A., Yocum, G.D., Robich, R.M. and Denlinger, D.L. (2007). Up-regulation of heat shock proteins is essential for col survival during insect diapause. *Proc. Natl. Acad .Sci. USA* 104, 1130-7.

Rutheford, S.L. and Lindquist, S. (1998). Hsp90 as a capacitor for morphological evolution. *Nature* 396, 336-342.

Rubin, D.M., Mehta, A.D., Zhu, J., Shoham, S., Chen, X., Wells, Q.R. and Palter, K.B. (1993). Genomic structure and sequence analysis of *Drosophila melanogaster* Hsc70 genes. *Gene* 128, 155-163.

Sakano, D., Li, B., Xia, Q., Yamamoto, K. (2006). Genes encoding small heat shock proteins of the silkworm, *Bombyx mori*. *Biosc. Biotechno. Biochem.* 70, 2443-2450.

Scheufler, C., Brinker, A., Bourenlov, G., Pegoraro, S., Moroder, L., Bartunik, H., Hartl, F.U. and Moarefi, I. (2000). Structure of TPR domain-peptide complexes: critical elements in the assembly of the HSP70-HSP90 multichaperone machine. *Cell* 101, 199-210.

Shue, G., and Kohtz, D.S. (1994). Structural and functional aspects of basic helix-loop-helix protein folding by heta-shock protein 90. *J Biol Chem* 269, 2707-2711.

Sonoda S., Fukumoto K., Izumi Y., Yoshida H., and Tsumuki H. (2006). Cloning of heat shock protein genes (hsp90 and hsc70) and their expression during larval diapause and cold tolerance acquisition in the rice stem borer *Chilo suppressalis* Walker. *Arc Insect Bioch Physiol* 63, 36-47.

Sonoda, S., Ashfaq, M. and Tsumuki, H. (2006). Cloning and nucleotide sequencing of three heat shock genes (hsp90, hsc70, and hsp19.5) from diamodback moth, *Plutella xylostella* (L.) and their expression in relation to developmental stage and temperature. *Arc. Insect Bioch. Physiol.* 62, 80-90.

Sonoda, S., Fukumoto, K., Izumi, Y., Ashfaq, M., Yoshida, H., Tsumuki, H. (2006). A small HSP gene is not responsible for diapause and cold tolerance acquisition in *Chilo suppressalis* . *Journal of Applied Entomology* 130, 309-313.

Spyliotopoulos, A., Gkouvitsas, T., Fantinou and A., Kourti, A. (2007). Expression of a cDNA encoding a member of the hexamerin storage proteins from the moth *Sesamia nonagrioides* (Lef.) during diapause. *Comp. Biochem. Physiol. B* 148, 44-54.

Tachibana, S.I., Numata, H. and Goto, S.G. (2005). Gene expression of heat-shock proteins (*Hsp23, Hsp70* and *Hsp90*) during and after larval diapause in the blow fly *Lucilia sericata*. *J Insect Physiol* 51, 641–647.

Tammariello, S.P., Denlinger, D.L. (1998). G_0/G_1 cell cycle arrest in the brain of *Sarcophaga crassipalpis* during pupal diapause and the expression pattern of the cell cycle regulator, proliferating cell nuclear antigen. *Insect Biochemical Molecular Biology* 28, 83–89.

Tauber, M.J., Tauber, C.A. and Masaki, S. (1986). Seasonal adaptations of insects. Oxford University Press, New York.

Thompson, J.D., Gibson, T.J., Plewniak, F., Jeanmougin, F. and Higgins, D.G. (1997). The ClustalX windows interface: flexible strategies for multiple sequence alignment aided by quality analysis tools. *Nucleic Acids Research* 25, 4876–4882.

Torchia, J., Rose, D.W., Inostroza, J., Kamei, Y., Westin, S., Glass, C.K., Rosenfeld, M.G. (1997). The transcriptional co-activator p/CIP binds CBP and mediates nuclear-receptor function. *Nature* 387, 677-684.

Tsitsipis, J.A. (1984). Rearing the corn borer *Sesamia nonagrioides* on artificial media in the laboratory. In Proceedings, XVII International Congress of Entomology, 16-26 August, Hamburg, Abstracts, p. 316.

van Montfort, R., Slingsby, C., Vierling, E. (2002). Structure and function of the small heat shock protein/α-crystallin family of molecular chaperones. *Adv. Protein Chem.* 59,105–156.

Walter, S., Buchner, J. (2002). Molecular chaperons-cellular machines for protein folding. *Angew Chem Int Ed Engl* 41, 1098-1113.

Whitesell, L, Mimnaugh , E.G., De Costa, B., Myers, C.E., and Neckers, L.M.. (1994). inhibition of heat shock protein HSP90-pp60v-src heteroprotein complex formation by benzoquinone ansamycins: essential role for stress proteins in ongogenic transformation. *Proc nat acad Sci USA* 91, 8324-8328.

Xu Y, and Lindquist, S. (1993). Heat-shock proteins 90 geverns the activity of pp60^{v-80rc} kinase. *Proc Natl Acad Sci USA* 90, 7074-7078.

Yiangou, M, Tsapogas, P., Nikolaidis, N. and Scouras, Z.G. (1997). Heat-shock gene expression during recovery after transient cold shock in *Drosophila auraria* (Diptera: Drosophillidae).*Cytobios* 92, 91-98.

Yin, C.M., Chippendale, G.M. (1979). Diapause of the southwestern corn borer, *Diatrea grandiosella*: further evidence showing juvenile hormone to be the regulator. *J Insect Physiol* 25, 513–523.

Yocum, G.D, Joplin, K.H, Denlinger, D.L.(1991). Expression of heat shock proteins in response to high and low temperature extremes in diapausing pharate larvae of the gypsy moth, *Lymantria dispar*. *Archives of Insect Biochemistry and Physiology* 18, 239–249.

Yocum, G.D., Joplin, K.H., Denlinger, D.L. (1998). Upregulation of a 23 kDa small heat shock protein transcript during pupal diapause in the flesh fly, *Sarcophaga crassipalpis*. *Insect Biochemistry and Molecular Biology* 28, 677-682.

Yocum, G.D. (2001). Differential expression of two HSP70 transcripts in response to cold shock, thermoperiod, and adult diapause in the Colorado potato beetle. *Jornal of Insect Physiology* 47, 1139–1145.

Yocum, G.D., Kemp, W.P., Bosch, J. and Knoblett, J.N.(2005). Temporal variation in overwintering gene expression and respiration in the solitary bee *Megachile rotundata*. *J. Insect. Physiol.* 51, 621-9.

Young, J. C., Moarefi, I. and Hartl, F.U. (2001). Hsp90: a specialized but essential protein-folding tool. *J Cell Biol* 154, 267-273.

In: Heat Shock Proteins
Editor: Saad Usmani
ISBN: 978-1-62417-571-8

Chapter VI

Reproductive Cycle and Temperature-Related Differences in Baseline Levels of HSP70 and Metallothioneins in Wild Oyster Populations of *Crassostrea Gigas*

Anne-Leïla Meistertzheim*
Laboratoire des Sciences de l'Environnement Marin (LEMAR), Institut Universitaire Européen de la Mer, Université de Bretagne Occidentale, Plouzané, France.
CEntre de Formation et de Recherche sur l'Environnement Méditerranéen (CEFREM), Université de Perpignan Via Domitia, Perpignan, France

Abstract

Heat Shock Protein 70 (HSP70) are generalist stress proteins expressed in response to numerous environmental stresses. However, the baselines of these constitutive proteins are still unknown in many organisms and in particular marine intertidal species such as the Pacific oyster, *Crassostrea gigas*. I demonstrate that levels of heat shock proteins (HSP) and other stress proteins (metallothioneins, MTs) quantified by ELISA, remained similar in gills, mantle and digestive gland between oysters inhabiting cold and hot sites.

In contrast, endogenous HSPs and MTs levels in gonad changed significantly during gametogenesis. In female gonads, the constitutive form of HSP70 and the MTs increased from immature to mature stages (about more than 3-fold) and decreased after spawning. In male gonads, the same expression patterns were observed, whereas protein levels were lower and decreased once fully mature. I hypothesize that the high level of stress proteins in eggs may increase survival of oyster progeny.

* Correspondence: A.L. Meistertzheim, CEntre de Formation et de Recherche sur l'Environnement Méditerranéen (CEFREM), UMR-CNRS 5110, Université de Perpignan, 58 Avenue Paul Alduy, 66860 Perpignan Cedex, France, Tel: +33 4 68 66 21 86, Fax: +33 4 68 66 22 81, E-mail: leila.meistertzheim@gmail.com.

Keywords: *Crassostrea gigas,* reproduction, Heat Shock Proteins, metallothioneins, sex, gonad, maturation stage, temperature, repartition area, thermal limit, ELISA

1. Introduction

Throughout their lives, all organisms must deal with changes in environmental and physiological parameters. They are almost constantly adjusting their metabolism, aiming to maintain homeostasis but sometimes requiring extreme responses in a bid to remain viable. Environmental challenges can include changes in oxygen, nutrient and water availability, salinity, temperature, atmospheric pressure, toxic metals, and UV radiation etc. The rocky intertidal zone presents some of the most challenging abiotic environmental conditions encountered by marine animals (Tomanek et Helmuth, 2002). In this habitat, environmental conditions range from fully aquatic to fully terrestrial over vertical distances of a few meters or less. Temperature and desiccation potential change seasonally and daily, depending on the tidal cycle and ambient weather conditions (Helmuth, 2002). During low tide, thermal stress and desiccation due to aerial emersion can affect growth, survival, and reproduction significantly (Blanchette et al., 2007).

Adaptations to temperature variations appear to be especially critical for survival in the intertidal zone (Somero, 2002). Once environmental temperatures start to approach the thermal limit of a given organism, its survival will depend on its capacity to effectively maintain or restore the integrity of the protein pool. One well-characterized cellular defense mechanism is the heat shock response (HSR), which involves the induction of a highly conserved group of molecular chaperones, also known as heat shock proteins (HSPs). Theses proteins are critical in the defence of protein homeostasis, the refolding of denatured proteins, and the breakdown and replacement of the proteins that are not repairable (see Tomanek (2008) for review). HSPs are now known to be a universal organism response to many kinds of stresses that can damage the proteome. Indeed, HSPs are an important and integral component of the minimal stress proteome of cells (Kültz, 2005). HSPs (and other elements of the stress response) also play roles in longevity and life extension (Krivoruchko et Storey, 2010; Kültz, 2005). HSPs fall into several major families based on their molecular masses, including Hsp100, Hsp90, Hsp70, Hsp60, Hsp40, and small HSPs (sHsps) (less than 30 kDa) families (Feder et Hofmann, 1999; Gething et Sambrook, 1992). It should be noted that a revised nomenclature that does not use a molecular mass designation has recently begun to be used in the mammalian literature and calls these families HSPH, HSPC, HSPA, HSPD, DNAJ, and HSPB, respectively (Kampinga et al., 2009).

Among HSP families, the most studied proteins are members of the 70-kDa heat shock protein (HSP70) family. HSP70 family is composed of both environmentally inducible (HSP) and constitutively expressed (HSC) family members (Clegg et al., 1998) that are encoded by distinct genes (Gourdon et al., 2000). The biosynthesis and chaperoning activities of HSPs are energetically costly, suggesting a tradeoff between thermal tolerance and use of metabolic energy for growth and reproduction (Somero, 2002). The production of heat shock proteins (HSP) is a passive and time-limited protection mechanism supported by anaerobic metabolism (Anestis et al., 2008), and is also an indicator of a species thermal sensitivity (reviewed by Feder and Hofmann (1999); Pörtner (2002)). It is commonly found that expression of inducible paralogs of HSPs occurs only above a certain threshold of induction temperature (Sanders et al., 1991; Hofmann and Somero, 1996; Feder and Hofmann, 1999; Tomanek and Somero, 2000). A number of studies have shown that animals inhabiting different microhabitats have dissimilar expression patterns of HSP70 (Anestis et al., 2008; Dong et al., 2008; Sanders et al., 1991; Tomanek et Somero, 1999), reflecting thermal adaptation to their local habitats. Functional significance of these proteins in natural populations of metazoans mainly came from work on *Drosophila* as well as that on intertidal ectotherms (Hoffmann et al., 2003; Meistertzheim et al., 2009; Tomanek, 2008). For example, differences in induction temperatures of HSPs in congeners of *Tegula* snails are based on both their vertical and latitudinal distribution patterns when exposed to heat stress in controlled conditions (Tomanek et Somero, 1999). Recently, the relationship between the vertical position of a species and the level of constitutive expression of heat-shock in the absence of thermal stress revealed distinct strategies of invertebrate species occupying different heights and orientations in the rocky intertidal zone (Dong, 2008). Two high-intertidal limpet species had higher constitutive levels of HSP70 than the low- and mid-intertidal species. However, little is known about the relationship between the biogeographic zonation of a species and the level of constitutive expression of heat-shock proteins at organismal and tissues levels in the absence of thermal stress and during the reproductive cycle (Meistertzheim et al., 2009).

To examine this relationship, I studied an invasive species, the Pacific cupped oysters *Crassostrea gigas* (Thunberg, 1789). This oyster is an important commercial species and has been successfully introduced into many parts of the world for aquaculture (Wolff et Reise, 2002). Introduced oysters reproduced successfully within several years and as a consequence, wild populations were found outside the culture plots. *C. gigas* is a protandric hermaphrodite species with a sex determinism more under genetic rather than environmental factors (Guo et al., 1998). Juveniles usually mature as male and could change to female later in life (Coe, 1943; Galtsoff, 1964). However, once gametogenesis has begun, oysters are unable to change sex until after they have spawned and the gonad is undifferentiated (Guo et al., 1998). The gonad of an oyster is not a discrete organ; rather, it is comprised of a network of tubules interspersed in connective tissue and surrounding the digestive gland (Lubet et al., 1976). Different germ cell developmental stages can occur simultaneously in the gonad including undifferentiated cells, growing germ cells, and mature gametes resulting in multiple cohorts of germ cells, each with synchronous development (Fabioux et al., 2004). Despite this, reproductive development is relatively synchronized during the year (Lango-Reynoso et al., 2000). Complex interactions between temperature and food quality and quantity affect gametogenesis of *C. gigas* (Chavez-Villalba et al., 2003). Reproductive development is temperature dependent in this species and occurs typically from February to September along the western Atlantic coast of France (Berthelin et al., 2000). Spawning occurs at a minimum seawater temperature of approximately 19°C (Mann, 1979).

Intertidal oysters are exposed to extreme physiological conditions during low tides (Newell, 1979) and experience water temperatures ranging from 4-24°C (Quayle, 1969). Furthermore, *C. gigas* body temperatures may exceed ambient air temperature and regularly approach sublethal thermal limits (43-44°C, 1h) (Clegg et al., 1998; Helmuth et Hofmann, 2001). Three isoforms of HSP70 family proteins were characterized in the oyster *C. gigas*. Two proteins, HSC77 and HSC72, are constitutively expressed, and their level of expression increases after acute thermal stress (Meistertzheim et al., 2007). In contrast to other bivalve species such as *Mytilus*, *C. gigas* can express an inducible form protein, HSP69, typically only detectable after acute heat or cold thermal stress (Airaksinen et al., 2003; Clegg et al., 1998). Several days after heat shock, the levels of HSP69 account for roughly one-third of total HSP70. In the absence of additional stress, HSP69 completely disappears within 7–10 days after heat shock (Clegg et al., 1998). In Pacific oysters, a single sublethal thermal shock is sufficient to cause significant HSP69 expression and also results in induced thermotolerance for a remarkably prolonged period of up to 2 weeks (Clegg et al., 1998). Thus, a great deal of plasticity in the threshold of induction temperature for HSPs was observed in oysters gill tissues as a function of season, tidal height (Halpin et al., 2002; Hamdoun et al., 2003; Roberts et al., 1997) and spawning (Li et al., 2007) when exposed to heat stress in controlled laboratory conditions. However, any variation of the levels of HPS70 was observed *in situ* on somatic tissues during gametogenesis in freshly field-collected specimens sampled at in low- and high-tidal heights (Meistertzheim et al., 2009). On the contrary, chaperones levels in gonadal tissues were influenced by sex and maturation stage and not by tidal heights. Similar patterns were also observed for other stress proteins, as Metallothioneins (MTs) (Meistertzheim et al., 2009). MTs are low-molecular weight, cysteine rich proteins with substantial metal binding capabilities. Metallothioneins play a role in the metabolism of the relatively non-toxic essential metals (zinc and copper) (Dunn, 1987) as well as in the detoxification of toxic metals such as cadmium (Viarengo et Nott, 1993). Three MT genes were identified in *Crassostrea gigas* and have been shown to be inducible by metallic stress (Tanguy et Moraga, 2001). During oxidative stress, synthesis of MTs may increase several-fold (Thornalley et Vasak, 1985) to protect the cells from cytotoxicity (Aschner et al., 1998) and DNA damage (Cai et al., 1995). Upregulation of MTs was also observed in response to cold stress in a marine gastropod *Littorina Littorea* (English et Storey, 2003).

Invasive populations of oysters on the west coast of Brittany provide an opportunity to examine baseline levels of both HSP70 and MTs during gametogenesis in both female and male in populations at the limits of species repartition area. The expression of soluble HSP70 and MTs was quantified in several organs of oysters from three different populations at same tidal height in conjunction with oyster sex and gametogenic stage. Oysters were chosen from non-contaminated sites in order to measure proteins levels in relation to reproductive development and spawning and thermal limit without the influence of anthropogenic stress.

2. Materials and Methods

2.1. Temperature Survey and Sampling

Crassostrea gigas invaded populations are reported in the entire French coast except in North-West coasts of the Brittany because of particular cold temperatures that prevent cultivated oysters spawning (Lejart, 2009). Three non-contaminated sites were chosen in cold, hot and intermediate zone in Brittany (Pointe de l'Arcouest: 48°47'57"N, Squiffiec: 48°22'26"N and Saint Pierre Quiberon (47°31'09"N) (Figure 1). Field temperatures were followed every 20 minutes from February to December 2005, using a temperature logger (EBI 85-A, Ebro) deployed inside oysters bag placed in oyster parks closed to sampling sites.

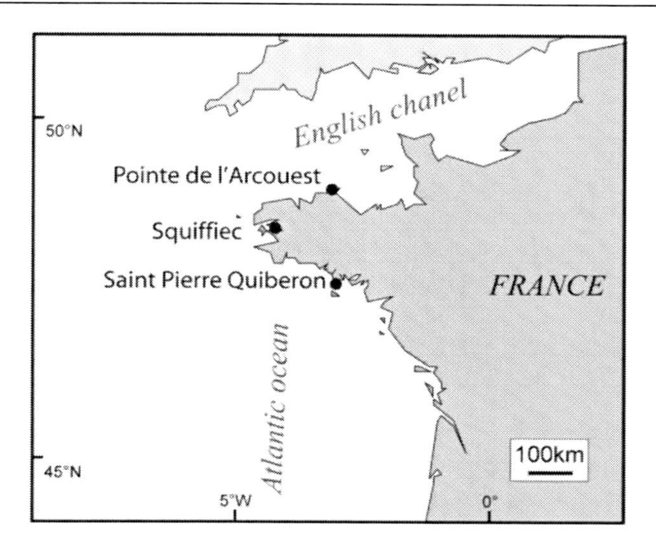

Figure 1. Geographical locations of the 3 French *Crassotrea gigas* populations.

Air and water temperatures were divided using vertical position of bags compared to water daily oscillations obtained from the Service Hydrographique et Océanographique de la Marine (SHOM). Temperatures are considered as similar to those encountered by oysters living at same tidal height on the rocky shore, closed to oyster parks. However, maximum air temperatures recorded at low-tidal during some summer afternoons could be lower than real temperatures expected in rocky areas due to the protection by oyster bags. Because of the large number of data points collected by the loggers, temperatures were summarized for each site on a monthly basis. The monthly average temperature and also the number of cumulate hours during which temperature was included in a given interval, were calculated including both aerial and submerged temperatures. Standard deviations of monthly average were recorded as a metric of variability between days within a month.

Adult oysters, *Crassostrea gigas*, were collected at same tidal height (corresponding to emersion times of 30%). Fifteen oysters were collected monthly in 2005, with two sampling per month during the gametogenic period (from April to September). For each animal, gills, mantle, gonad and digestive gland were dissected, frozen in liquid nitrogen and stored at -80°C until analyses; a 3-mm cross section of the visceral mass was also excised for histological examination as described below.

2.2. Histological Analysis

A standard section of the visceral mass was taken at the intersection of the mantle and gonad, fixed in Bouin's fluid for 24h. Tissues were dehydrated through an ascending ethanol series and embedded in paraffin. Deparaffinized 5-μm-thick sections were stained according to the trichrome protocol (Gabe, 1968) and examined under a light microscope to determine sex and gametogenic activity. Morphological condition of the gonad was classified according to the modified scheme proposed by Lubet (1959): stage 0, resting stage, stage I, multiplication of gonia, stage II, progression of gametogenesis, stage IIIA, ripe gonad; stage IIID, reabsorbing stage, IV post-spawning (Table 1).

Table 1. Reproductive scale for *Crassostrea gigas* according to the modified scheme proposed by Lubet (1959)

Stage	Histological description
0-	Sexual rest. Indeterminate sex. Rare acini with germinal cells
I	Proliferation of gonia and number of acini.
II	Beginning of maturation of sexual cells. Some mature gametes.
III	Follicles completely filled of mature gametes (relatively homogeneous size)
IV	Follicles containing degenerating gametes. Oocytes often elongated in shape sometimes broken. Obvious redevelopment indicated by increased number of primary oocytes.
V	Rest post-spawning. Indeterminate sex.

2.3. Quantification of HSP70, Mts and Eggs by ELISA

Sex and gametogenic stage were identified in all samples. Five oysters from each sampling date, in each site were grouped by sex and stage. For each category, samples of gills, mantle, gonads and digestive glands were collected, homogenized in protein extraction buffer (150 mM NaCl, 10 mM NaH_2PO_4, 1 mM phenylmethanesulfonyl fluoride pH = 7.2) and centrifuged. Protein concentration was estimated with a Dc Protein Assay kit (Bio-Rad, Hercules, CA, USA) using BSA as the standard. Optical density was measured at 620 nm using a microplate reader. Microtiter plates were coated with 20 µg $well^{-1}$ of total proteins and incubated over night at 4 °C. HSP70 and MTs concentrations were estimated by ELISA using rabbit anti-*Cg*Hsc72 and anti-*Cg*Mt biotinylated polyclonal antibodies and recombinant *Cg*Hsc72 and *Cg*Mt proteins, respectively, as standards, according to procedures previously described (Boutet et al., 2002; Boutet et al., 2003). Because the gonad tissue develops between the digestive gland and mantle during gametogenesis, it was not possible to remove all the gonad tissue from the digestive gland. Egg protein levels were quantified and considered as a "contamination" in the digestive gland. Polyclonal antibodies developed from egg proteins of *C. gigas* (Kang et al., 2003) were used on the same samples to estimate the presence of oocytes in digestive gland samples according to the procedure described by the authors. To create a standard curve, proteins were extracted from purified oyster eggs and 10-fold serial dilutions and assayed in triplicate. HSP70, MT and egg proteins were quantified for each individual from gonad and digestive gland tissue samples. Corrections were made for HSP70 and MTs levels of the digestive gland using HSP70 and MTs levels measured in oocytes and percentage of oocytes in digestive gland.

2.4. Western Blotting Immunoassay for HSP70 Family

Western blots of oysters HSP70 family were prepared according to Clegg et al. (1998). Briefly, aliquots (10µg) of each solubilized sample were resolved in duplicate on a 10% SDS-polyacrylamide gel. To normalize between blots, an aliquot of a "reference sample" of heat-shocked oyster gill was added to the first and last lanes of each gel (soluble protein obtained by heat shocking four oysters for 1h at 37°C followed by a 48h accumulation period at ambient temperature). This reference was known to contain isoforms HSP72 and HSP69 of the HSP70 family. Blotting procedures consisted of transferring proteins onto nitrocellulose (0.45 μm). The blot was blocked with 1% Bovine Serum Albumin in TTBS (0.05% Tween-20, 500 mM NaCl, 15 mM Tris, pH 7.5) overnight at 4°C, followed by two washes in TBS (500 mM NaCl, 15 mM Tris, pH 7.5) for 10 min. Blots were cut in the center lane and two different blots were prepared simultaneously. The left half of each blot was probed with Affinity BioReagents monoclonal anti-hsp-70, clone 7.10 (catalogue number MA3-006, Golden, Colo.), to recognize HSC72 and HSP69. The right half of each blot was probed with rabbit anti-*Cg*Hsc72 polyclonal antibodies previously used in the ELISA assay. Antibody dilutions (1:3000 primary antibody applied for 2h and 1:1000 secondary antibody applied for 1.5h) used were selected to ensure that the quantity of antigen, not antibody, was limiting. Secondary antibodies consisted of mouse-antihuman IgG and goat-antirabbit IgG, respectively, conjugated to horseradish peroxidase (Sigma Chemical Co., St. Louis, Mo.). Labeled proteins were detected with ECL Western blotting reagents (Amersham Corp., Arlington Heights, Ill.) using Kodak BioMax MR single emulsion film (Eastman Kodak Co., Rochester, N.Y.). The size of individual protein bands were estimated by regression against a standard curve of known molecular weight proteins.

2.5. Statistical Analysis

The data obtained were not normally distributed. Differences in stress proteins expression between oysters from the three sites were tested using a Kruskal-Wallis ANOVA at each gametogenic stage. Differences between gametogenic stages for oysters at the same site were identified using a Kruskal-Wallis ANOVA and a sequential Bonferroni correction of the p-value. All statistical analyses were carried out with Statistica version 10 Software (Statsoft) at a significance level of α=0.05.

3. Results

3.1 Temperature Survey

The mean water and total temperatures (air and water) were higher in Saint Pierre Quiberon than in Squiffiec which were higher than Pointe de l'Arcouest (Table 2). Mean annual water temperatures ranged from 14.5°C (warmest site) to 13.8°C (coldest site). The time during which temperature exceeds the threshold temperature of 19°C in the three sites, corresponded to 58.1, 28.9 and 36 days (cumulated hours) in water and to 69.8, 39.6 and 40.9 days in total (air and water taking together) for Saint Pierre Quiberon, Squiffiec and Pointe de l'Arcouest, respectively. Mean water temperature between Squiffiec and Pointe de l'Arcouest were relatively similar in summer, but temperature profiles differ with 11.3 days of total temperatures higher than 20°C in Squiffiec compared to 7.5 d in Pointe de l'Arcouest (data not shown).

Table 2. Cumulate hours t (days) of temperatures ranged in different intervals measured in water (w) and total (tot, air and water) by temperature loggers during 2005 placed in oysters bag, closed to the sites. Monthly mean water temperature is also indicated with standard deviation the sites of Pointe de l'Arcouest, Squiffiec and Saint-Pierre Quiberon (Brittany, France)

Date	Pointe de l'Arcouest							Squiffiec							St Pierre Quiberon						
	$t_{T°C<15}$		$t_{15<T°C<19}$		$t_{T°C>19}$		T°C (SD)	$t_{T°C<15}$		$t_{15<T°C<19}$		$t_{T°C>19}$		T°C (SD)	$t_{T°C<15}$		$t_{15<T°C<19}$		$t_{T°C>19}$		T°C (SD)
	w	*tot*	*w*	*tot*	*w*	*tot*	*w*	*w*	*tot*	*w*	*tot*	*w*	*tot*	*w*	*w*	*tot*	*w*	*tot*	*w*	*tot*	*w*
Feb	15.0	18.0					7.6 (0.9)	11.4	15.6					9.3 (0.3)	11.7	18.6					7.4 (1.0)
Mar	28	30.8					7.7 (1.5)	22.0	31.0		0.2			8.1 (0.8)	23.1	30.9		0.07			9.2 (3.1)
Apr	25.7	29.1					10.4 (0.5)	21.0	29.2		1.0			11.5 (0.8)	18.1	24.6	0.1	1.0			12.3 (1.0)
May	25.1	28.8		0.6			12.9 (0.7)	21.6	30.0		18.4		0.7	12.9 (0.8)	9.4	12.3	13.9	17.7	0.4	1.0	15.5 (1.5)
June	6.7	7.8	17.3	21.4		0.1	15.9 (1.1)	7.3	11.0	14.4	11.5		19.2	15.8 (1.1)		0.3	11.5	12.6	13.8	17.1	19.2 (1.7)
July			22.3	23.7	5.8	7.3	18.3 (0.9)		0.3	8.4	16.2	14.3	13.5	19.3 (0.7)		0.2	12.2	13.4	14.1	17.4	19.5 (1.3)
Aug		0.1	11.0	12.6	15.4	18.3	19.2 (0.5)		1.3	13.2	22.1	10.3	6.0	18.9 (0.5)		0.6	10.1	11.5	15.8	19.0	19.3 (0.8)
Sep		0.8	13.0	14.0	14.8	15.2	18.8 (0.8)		1.9	19.5	27.3	4.3	0.2	18.1 (0.8)		1.0	12.5	13.7	14.0	15.3	19.2 (1.2)
Oct		1.0	29.2	30.0			16.6 (0.4)	0.3	3.4	24.6	1.1			16.0 (0.4)	0.3	2.2	27.5	28.8			16.4 (0.5)
Nov	21.0	24.8	4.6	4.6			12.8 (1.9)	22.2	27.8	0.82				12.2 (2.1)	22.9	27.4	2.5	2.5			11.3 (2.9)
Dec	28.4	29.8					9.1 (0.9)	5.5	8.9					8.2 (1.9)	27.5	30.8					7.1 (1.5)
Total	149.9	171.0	97.4	106.9	36.0	40.9	13.8 (4.3)	111.3	160.4	80.9	97.8	28.9	39.6	14.3 (3.9)	113.9	148.9	90.3	101.3	58.1	69.8	14.5 (4.9)

3.2. Quantification of HSP70 and MTs by ELISA

HSP70 and MT protein levels measured by ELISA appeared to be tissue dependent and showed similar patterns in oysters sampled in the three sites (Figure 2-3). Few significant differences were observed between oysters groups within the same gametogenic stage. HSP70 levels in digestive gland were higher in females stage I and II, and MTs levels in stage V from Pointe de l'Arcouest than the both others sites ($p<0.05$). MTs levels were significantly higher in gonad from female oysters at stage III sampled from Pointe de l'Arcouest than those from Saint Pierre Quiberon ($p<0.011$).

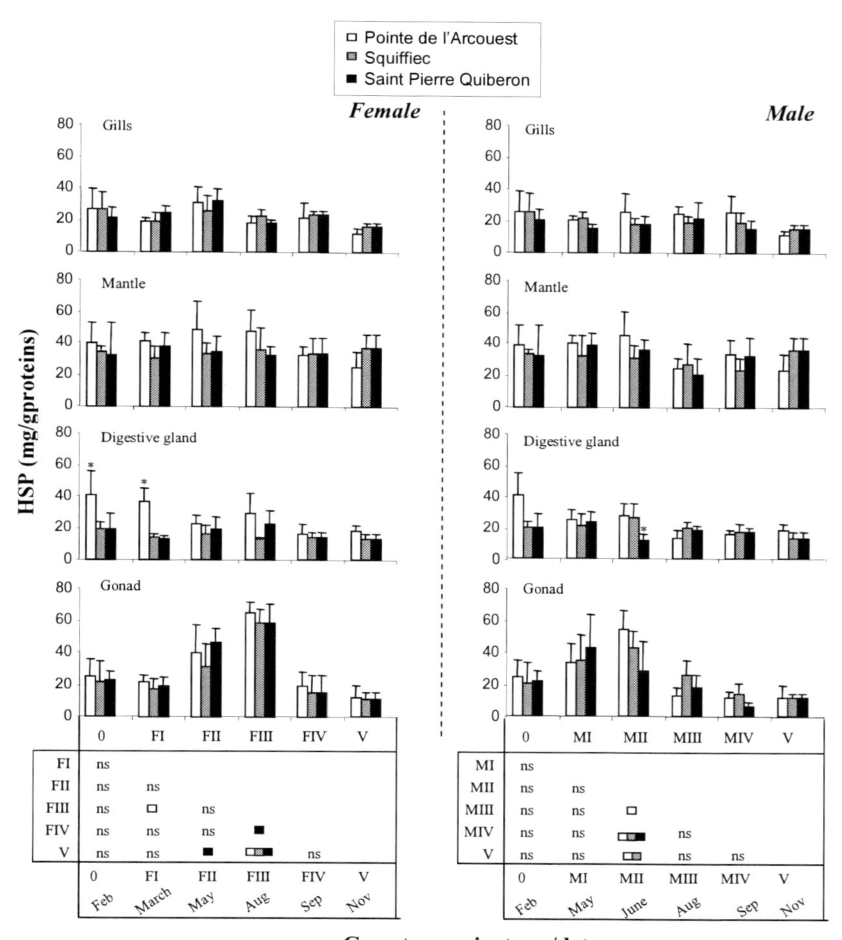

Gametogenesis stage /date

* Significant differences at $p<0.05$ between oyster from the three sites at the same gametogenic stage.

Significant differences at $p<0.05$ after sequential corrections of Bonferroni between gonad reproductive stages per site.

Figure 2. Quantification of HSP70 in *Crassostrea gigas* during the gametogenesis in females (F) and males (M) in three sites (Pointe de l'Arcouest, Squiffiec and Saint Pierre Quiberon) for each gametogenic stage: sex-indeterminate at stage 0 (resting stage, indicated by 0); stage I, multiplication of gonia; stage II, progression of gametogenesis; stage III, ripe gonad; stage IV, reabsorbing stage and stage V, restoring stage (sex-indeterminate). Bars represent the mean of five individuals per sampling point within each gametogenic stage; error bars correspond to standard deviation.

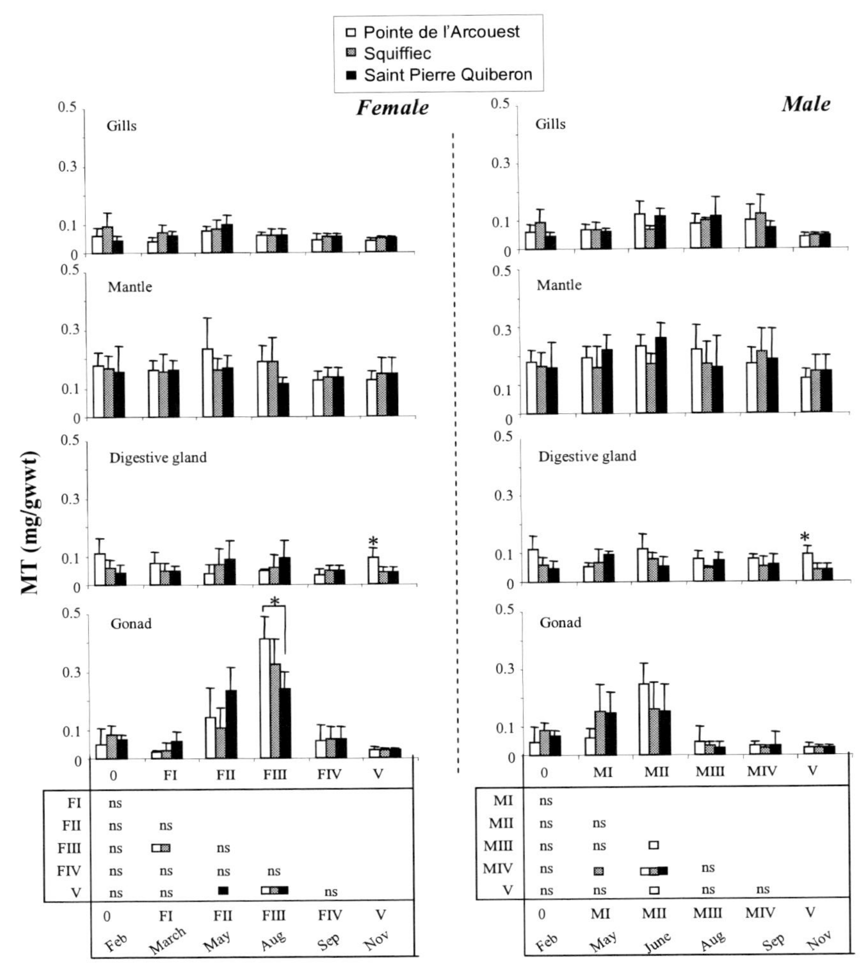

* Significant differences at p<0.05 between oyster from the three sites at the same gametogenic stage.

Significant differences at p<0.05 after sequential corrections of Bonferroni between gonad reproductive stages per site.

Figure 3. Quantification of MTs in *Crassostrea gigas* during the gametogenesis in females (F) and males (M) in three sites (Pointe de l'Arcouest, Squiffiec and Saint Pierre Quiberon) for each gametogenic stage: sex-indeterminate at stage 0 (resting stage, indicated by 0); stage I, multiplication of gonia; stage II, progression of gametogenesis; stage III, ripe gonad; stage IV, reabsorbing stage and stage V, restoring stage (sex-indeterminate). Bars represent the mean of five individuals per sampling point within each gametogenic stage; error bars correspond to standard deviation.

No significant difference in baseline levels were detected between oysters sampled during winter (stage 0) and summer (stage III) in mantle and gills tissues ($p>0.05$).

Significant differences in HSP70 and MTs were observed in gonadal tissues of oysters from the three sites between sex and gametogenic stages (Figure 2-3). The highest HSP70 levels were obtained from females with ripe gonad tissue (stage III; $p<0.01$) and corresponded to approximately 6% of the total amount proteins extracted from gonad tissue. Conversely, the lowest HSP70 levels were obtained in post-spawning individuals (stage V-indeterminate sex). Maximum levels of HSP70 were lower in males than in females ($p<0.04$).

Levels of HSP70 measured increased in females from stage 0 (rest) to the stage III (ripe) and decreased from stage IV (restoring stage) to stage V (post-spawning). For males, the HSP70 levels increased significantly from stage 0 until stage II (growing) and decreased from stage III to the stage V (Figure 2). A similar pattern was observed for MTs levels in both males and females (Figure 3). The highest MTs levels were obtained in females, in the ripe gonad (stage III) corresponding to 0.7% of the total amount proteins.

3.3. Identification of HSP70 Isoforms Quantified by ELISA

The isoforms of HSP70 family measured by ELISA were identified by western blot as HSC72 in comparison with the reference sample that contained HSC72 and HSP69. The constitutive form HSC72 was identified in female and male gonad, gills and digestive gland at stage 0 and III for each oyster population (Figure 4).

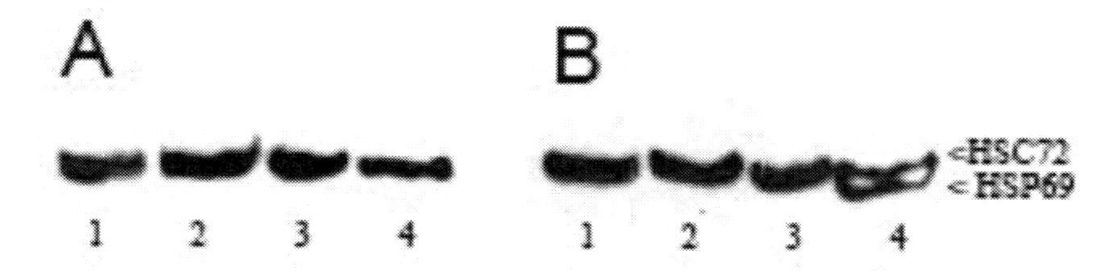

Figure 4. Heat shock protein expression in oyster tissues at varying reproductive stages using polyclonal rabbit anti-*Cg*Hsc72 antibody (A) and a commercial monoclonal mouse anti-HSP70 (B) as described in Meistertzheim et al. (2009). Lanes 1 and 2 contain digestive gland tissue in oysters with a resting gonad and mature female gonad, respectively. Lane 3 represents gonadal tissue from the same oyster in Lane 2. Lane 4 represents control (gill tissue) of a 37°C heat-shocked oyster. Molecular mass is shown on the right.

4. Discussion

4.1. Expression Patterns of HSP70 and MTs in Gonads during Gametogenesis

In the present study, the similar induction of the constitutive form HSP70 (HSC72) and MTs in *C. gigas* gonadal tissues suggests an important role for these stress proteins during reproduction. In males, cognate HSP70 and MT levels increased continuously during proliferation phase of spermatogonia with a maxima in developing but not ripe stages. In females, levels continued to increase until maturation of oocytes, which are naturally stopped at the first division of meiosis, and decreased quickly after spawning. When totally mature, the female gonad is almost entirely composed of oocytes at more than 90% (Lubet, 1959), suggesting the presence of HSP70 and MT mainly in oocytes compared to other gonad associated tissue (*e.g.* gonad epithelium that is a larger component of immature gonad). Maturing oocytes contained 41% of total proteins (Kang et al., 2003), of which cognate HSP70 and MTs levels represented 6% and 0.7%, respectively. The essential role of constitutive HSP70 molecular chaperone has been previously determined for sea urchin oocytes (Geraci et al., 2003) and mice germ cells meiosis (Eddy, 1998). The high levels of MTs in female and male gonads detected in this study suggest also a role for MTs during meiosis (Olesen et al., 2004).

MTs have been previously shown to be localized in the nucleus of oyster oocytes (Moraga et al., 2001) and in oyster embryos (Ringwood et Brouwer, 1995). Cell division is a key element in the development of embryos and also of some diseases such as various cancers. MTs play a significant role in mechanisms which control growth, differentiation and proliferation of human cells (see Dziegiel (2004) for review). In addition, the essential role of constitutive HSP70 molecular chaperone has been also previously determined for mitosis during early embryo cleavage of undifferentiated cells (Sconzo et al., 1999), maintaining appropriate cell numbers, that depends on the balance between proliferation, differentiation and apoptosis (reviewed in Rupik et al. (2011)). High constitutive HSP70 and MTs levels in oyster oocytes in this study suggest a protective role also in mitosis at first stages of embryo-larval cell division.

MTs are also involved in cellular protection from actions of harmful agents (metals, free radicals, etc.). High levels of MTs in oocytes could increase the chance of larval survival in a high variable environment. Early embryo-larval stages of bivalve mollusks have been shown to be highly sensitive to micropollutants, particularly to metals (Calabrese et al., 1977; Martin et al., 1981). The importance of MTs in ameliorating protection against metal toxicity in bivalves has been well demonstrated in embryo-larval stages of mollusks (Roesijadi et al., 1996) and also in adults (Roesijadi, 1992). MT induction in *C. gigas* larvae exposed to metal contamination only occurs when gametes are obtained from parents at maximum sexual maturity, *e.g.* when gametes reach maximal size (Damiens et al., 2006). In addition, induction of HSP70 in American horseshoe crab, *Limulus polyphemus* L., larvae occurs a few hours after heat shock and is correlated to high larvae survival; however, this stress protein was not similarly induced in embryos (Botton et al., 2006). Similar HSP70 induction profile was observed in vertebrates heat-shocked oocytes (Mortensen et al., 2010). Therefore during oogenesis, maternal mRNA and proteins are transported from follicle cells into the oocytes and affect embryo development (Anderson et Nusslein-Volhard, 1984). In this study, high levels proteins in oocytes in the three sites suggest that constitutive HSP70 and MTs in embryos are provided via maternal transmission. The high baseline of these proteins in *C. gigas* oocytes may be an effective strategy to respond to environmental stress encountered by benthic, negatively buoyant embryos and pelagic larvae of these intertidal organisms.

4.2. Seasonal and Geographical Variations of Stress Proteins in Somatic Tissues

Sessile intertidal invertebrates are of particular interest, because they are often exposed to extreme environmental fluctuations, including seasonal and tidal cyclic temperature changes during periods of emersion and immersion (Helmuth, 1999). Physiological stressors such as osmotic stress, hypoxia and desiccation may contribute to the influence of temperature on HSPs induction in intertidal species (Feder et Hofmann, 1999; Greene et al., 2011). In *C. gigas,* higher levels of inducible HSP70 were measured in winter than in summer in gills of cultured individuals and were positively correlated to tidal height (Hamdoun et al., 2003).

However, several factors due to culture practices, such as hypoxia, high density of pathogens and proximity to sediment, are typically strong stressors (Burge et al., 2007; Soletchnik et al., 2005) and could induce HSP70 (David et al., 2005; Encomio et Chu, 2005). *In situ*, bivalves HSP70 expression in response to seasonal thermal changes were followed over the course of the year in several bivalves (English et Storey, 2003; Hamdoun et al., 2003; Meistertzheim et al., 2009). For example, inducible form of HSP70 in gills mussels (*Mytilus edulis* and *M. trossolus*) showed higher levels in summer than in winter (Chapple et al., 1998; Hofmann et Somero, 1995). However, no significant seasonal variation of both inducible and constitutive forms of HSP70 was observed at different tidal heights in mussels, *M. californianus* (Helmuth et Hofmann, 2001). In this study, no seasonal variation was observed in the stress proteins levels in different somatic tissues of wild oysters (gills, mantle and digestive gland), as previously observed in individuals sampled at two different tidal heights in Squiffiec (Meistertzheim et al., 2009). Only few differences in stress proteins levels between populations of the coldest and the warmest sites were measured in the digestive gland. Constitutive HSP70 and MTs levels were higher in females sampled in Pointe de l'Arcouest compared to Saint Pierre Quiberon during the first stages of the gametogenesis, corresponding to February and March 2005. Interestingly, water temperature was colder in Pointe de l'Arcouest than in Saint Pierre Quiberon ($t_{T°C<15}$ of 83 % and 63% of the time, respectively). High levels of constitutive HSP70 and MTs in the coldest site could be due to particular cold stress as observed in other marine species (Airaksinen et al., 2003; English et Storey, 2003). Organisms' HSPs proteins baseline levels could reflected thermal tolerance potential and is essential to understand how they are adapted to their current habitats and whether some species are living closer to their physiological limits than others (Hochachka et Somero, 2002; Somero, 2010). In this study, stress response revealed by high levels of constitutive form of HSP70 during winter in the coldest site could explain the limit of *C. gigas* repartition in the area.

In oysters, constitutive HSP70 and MTs proteins are also induced in responses to different contamination in controlled conditions (Amiard et al., 2006; Boutet et al., 2004; Boutet et al., 2003; Damiens et al., 2006) or hypoxia (David et al., 2005). Generally, theses chaperone proteins are used as biomarkers of the oxidative stress (David et al., 2005; Valavanidis et al., 2006). Here, absence of strong variation in the baseline levels of MTs and constitutive HSP70 during the year at different temperatures could suggest an absence of strong stress during 2005 in wild oyster's populations. Absence of the inducible form of HSP70 (HSP69) that reflects an acute stress-inducible subset of the 70-kDa HSPs reinforces this hypothesis.

4.3. Biomarkers and Reproduction

Finally, high levels of MTs and constitutive HPS70 were only correlated with the presence of gametes suggesting a protective role for progeny. Its important to remember that oyster spawning in summer is triggered by elevated temperature and is associated with the expenditure of energy (Berthelin et al., 2000; Mao et al., 2006). Inducible form of HSP70 can be induced after a heat shock in gills of adult oysters before spawning and has been correlated to an increase of thermotolerance before but not after spawning (Li et al., 2007). The energy expended during gametogenesis appears to compromise their thermotolerance, leaving oysters easily subjected to mortality if stress occurs in post-spawning. In contrast, high constitutive form of HSP70 and MTs levels in oyster oocytes could increase the potential for larval survival and, consequently, of this species by protecting the early embryo-larval stages from stress (Hamdoun et Epel, 2007). Despite the energetic demand of stress protein production, high levels of HSPs and MTs in oocytes likely increase the ability of developing embryos and larvae to survive in stressful environments, and therefore would constitute a strong adaptation strategy for this species. To conclude, the "inducible" form of HSP70 appeared as a better biomarker to study response to heat stress and thermal tolerance than the "constitutive" form, clearly affected by multiple physiological parameters such as gametogenesis. Although the constitutive form of HSP70 does not necessarily change with thermal stress, this form gives additional information on the adaptation strategy of species.

Acknowledgments

I gratefully acknowledge Marie-Thérèse Thébault for her helpful suggestions, comments and corrections during the study. I am grateful to A. Marhic for assistance with sample processing and to L. Coïc and C. Bertolone for technical support in the histological study and T. Lagadec in sensor data analysis. I also thank C. Seguineau and Kyung-Il Park for supplying of the polyclonal antibody developed from egg proteins and D. Moraga for the antibody against HSC72 and MTs of *Crassostrea gigas*. This research program was financially supported by the national program PROGIG (Prolifération de *Crassostrea gigas,* LITEAU II).

Reviewed by Marie-Thérèse Thébault and Elisabeth Faliex

References

Airaksinen, S., Jokilehto, T., Rabergh, C.M.I., Nikinmaa, M., 2003. Heat- and cold-inducible regulation of HSP70 expression in zebrafish ZF4 cells. *Comparative Biochemistry and Physiology Part B: Biochemistry and Molecular Biology* 136, 275-282.

Amiard, J.C., Amiard-Triquet, C., Barka, S., Pellerin, J., Rainbow, P.S., 2006. Metallothioneins in aquatic invertebrates: Their role in metal detoxification and their use as biomarkers. *Aquatic Toxicology* 76, 160-202.

Anderson, K.V., Nusslein-Volhard, C., 1984. Information for the dorsal-ventral pattern of the Drosophila embryo is stored as maternal mRNA. *Nature* 311, 223-227.

Anestis, A., Pörtner, H.O., Lazou, A., Michaelidis, B., 2008. Metabolic and molecular stress responses of sublittoral bearded horse mussel *Modiolus barbatus* to warming sea water: implications for vertical zonation. *Journal of Experimental Biology* 211, 2889-2898.

Aschner, M., Conklin, D.R., Yao, C.P., Allen, J.W., Tan, K.H., 1998. Induction of astrocyte metallothioneins (MTs) by zinc confers resistance against the acute cytotoxic effects of methylmercury on cell swelling, Na+ uptake, and K+ release. *Brain Research* 813, 254-261.

Berthelin, C., Kellner, K., Mathieu, M., 2000. Storage metabolism in the Pacific oyster (*Crassostrea gigas*) in relation to summer mortalities and reproductive cycle (west coast of France). *Comparative Biochemistry and Physiology. Part B, Biochemistry and Molecular Biology* 125, 359-369.

Blanchette, C.A., Helmuth, B., Gaines, S.D., 2007. Spatial patterns of growth in the mussel, *Mytilus californianus*, across a major oceanographic and biogeographic boundary at Point Conception, California, USA. *Journal of Experimental Marine Biology and Ecology* 340, 126-148.

Botton, M.L., Pogorzelska, M., Smoral, L., Shehata, A., Hamilton, M.G., 2006. Thermal biology of horseshoe crab embryos and larvae: A role for heat shock proteins. *Journal of Experimental Marine Biology and Ecology* 336, 65-73.

Boutet, I., Tanguy, A., Moraga, D., 2004. Response of the Pacific oyster *Crassostrea gigas* to hydrocarbon contamination under experimental conditions. *Gene* 329, 147-157.

Boutet, I., Tanguy, A., Auffret, M., Riso, R., Moraga, D., 2002. Immunochemical quantification of metallothioneins in marine mollusks: characterization of a metal exposure bioindicator. *Environmental Toxicology and Chemistry* 21, 1009-1014.

Boutet, I., Tanguy, A., Rousseau, S., Auffret, M., Moraga, D., 2003. Molecular identification and expression of heat shock cognate 70 (hsc70) and heat shock protein 70 (hsp70) genes in the Pacific oyster *Crassostrea gigas*. *Cell Stress & Chaperones* 8, 76-85.

Burge, C.A., Judah, L.R., Conquest, L.L., Griffin, F.J., Cheney, D.P., Suhrbier, A., Vadopalas, B., Olin, P.G., Renault, T., Friedman, C.S., 2007. Summer seed mortality of the Pacific oyster, *Crassostrea gigas* (Thunberg) grown in Tomales Bay, California, USA: The influence of oyster stock, planting time, pathogens, and environmental stressors. *Journal of Shellfish Research* 26, 163-172.

Cai, L., Koropatnick, J., Cherian, M.G., 1995. Metallothionein protects DNA from copper-induced but not iron-induced cleavage in vitro. *Chemico-Biological Interactions* 96, 143-155.

Calabrese, A., MacInnes, J.R., Nelson, D.A., Miller, J.E., 1977. Survival and growth of bivalve larvae under heavy-metal stress. *Marine Biology* 41, 179-184.

Chapple, J.P., Smerdon, G.R., J., B.R., S., H.J., 1998. Seasonal changes in stress-70 protein levels reflect thermal tolerance in the marine bivalve *Mytilus edulis*. *Journal of Experimental Marine Biology and Ecology* 229, 53-68.

Chavez-Villalba, J., Cochard, J.C., Pennec, M., Barret, J., Enriquez-Diaz, M., Caceres-Martinez, C., 2003. Effects of temperature and feeding regimes on gametogenesis and larval production in the oyster *Crassostrea gigas*. *Journal of Shellfish Research* 22, 721-731.

Clegg, J.S., Uhlinger, K.R., Jackson, S.A., Cherr, G.N., Rifkin, E., Friedman, C.S., 1998. Induced thermotolerance and the heat shock protein–70 family in the Pacific oyster *Crassostrea gigas*. *Molecular Marine Biology and Biothechnology* 7, 21-30.

Coe, W.R., 1943. Sexual differentiation in molluscs. *Quarterly Review of Biology* 18, 154-164.

Damiens, G., Mouneyrac, C., Quiniou, F., His, E., Gnassia-Barelli, M., Roméo, M., 2006. Metal bioaccumulation and metallothionein concentrations in larvae of *Crassostrea gigas*. *Environmental Pollution* 140, 492-499.

David, E., Tanguy, A., Pichavant, K., Moraga, D., 2005. Response of the Pacific oyster *Crassostrea gigas* to hypoxia exposure under experimental conditions. *FEBS Journal* 272, 5635-5652.

Dong, Y., Miller, L.P., Sanders, J.G., Somero, G.N., 2008. Heat-Shock Protein 70 (Hsp70) expression in four limpets of the genus *Lottia*: interspecific variation in constitutive and inducible synthesis correlates with *in situ* exposure to heat stress. *The Biological Bulletin* 215, 173-181.

Dunn, M.A., 1987. Metallothionein. *Proceedings of the Society for Experimental Biology and Medicine* 185, 107-119.

Dziegiel, P., 2004. Expression of metallothioneins in tumor cells. *Polish Journal of Pathology* 55, 3.

Eddy, E.M., 1998. HSP70-2 heat-shock protein of mouse spermatogenic cells. *Journal of Experimental Zoology* 282, 261-271.

Encomio, V.G., Chu, F.L.E., 2005. Seasonal variation of Heat Shock Protein 70 in eastern oysters (*Crassostrea virginica*) infected with *Perkinsus marinus* (Dermo). *Journal of Shellfish Research* 24, 167-175.

English, T.E., Storey, K.B., 2003. Freezing and anoxia stresses induce expression of metallothionein in the foot muscle and hepatopancreas of the marine gastropod *Littorina littorea*. *The Journal of Experimental Biology* 206, 2517-2524.

Fabioux, C., Pouvreau, S., Le Roux, F., Huvet, A., 2004. The oyster vasa-like gene: a specific marker of the germline in *Crassostrea gigas*. *Biochemical and Biophysical Research Communications* 315, 897-904.

Feder, M.E., Hofmann, G.E., 1999. Heat-shock proteins, molecular chaperones, and the stress response: evolutionary and ecological physiology. *Annual Review of Physiology* 61, 243-282.

Gabe, M., 1968. Techniques histologiques, Paris, 1113 p. pp.

Galtsoff, P.S., 1964. The American oyster *Crassostrea virginica* Gmelin. *Fish Bulletin* 64, 1-480.

Geraci, F., Agueli, C., Giudice, G., Sconzo, G., 2003. Localization of HSP70, Cdc2, and cyclin B in sea urchin oocytes in non-stressed conditions. *Biochemical and Biophysical Research Communications* 310, 748-753.

Gething, M.-J., Sambrook, J., 1992. Protein folding in the cell. *Nature* 355, 33-45.

Gourdon, I., Gricourt, L., Kellner, K., Roch, P., Escoubas, J.M., 2000. Characterization of a cDNA encoding a 72 kDa heat shock cognate protein (Hsc72) from the Pacific oyster, *Crassostrea gigas*. *DNA Sequence* 11, 265-270.

Greene, M., Hamilton, M., Botton, M., 2011. Physiological responses of horseshoe crab (*Limulus polyphemus*) embryos to osmotic stress and a possible role for stress proteins (HSPs). *Marine Biology* 158, 1691-1698.

Guo, X., Hedgecock, D., Hershberger, W.K., Cooper, K., Allen Jr, S.K., 1998. Genetic Determinants of Protandric Sex in the Pacific Oyster, *Crassostrea gigas* Thunberg. *Evolution* 52, 394-402.

Halpin, P.M., Sorte, C.J., Hofmann, G.E., Menge, B.A., 2002. Patterns of Variation in Levels of Hsp70 in Natural Rocky Shore Populations from Microscales to Mesoscales. *Integrative and Comparative Biology* 42, 815-824.

Hamdoun, A., Epel, D., 2007. Embryo stability and vulnerability in an always changing world. *Proceedings of the National Academy of Sciences of United States of America* 104, 1745.

Hamdoun, A.M., Cheney, D.P., Cherr, G.N., 2003. Phenotypic plasticity of HSP70 and HSP70 gene expression in the Pacific oyster (*Crassostrea gigas*): implications for thermal limits and induction of thermal tolerance. *Biological Bulletin* 205, 160-169.

Helmuth, B., 1999. Thermal Biology of Rocky Intertidal Mussels: Quantifying Body Temperatures Using Climatological Data. *Ecology* 80, 15-34.

Helmuth, B., 2002. How do we measure the environment? Linking intertidal thermal physiology and ecology through biophysics. *Integrative and Comparative Biology* 42, 837-845.

Helmuth, B.S., Hofmann, G.E., 2001. Microhabitats, thermal heterogeneity, and patterns of physiological stress in the rocky intertidal zone. *Biology Bulletin* 201, 374-384.

Hochachka, P.W., Somero, G.N., 2002. Biochemical adaptation: mechanism and process in physiological evolution. Oxford University Press, New York, 480 pp. pp.

Hoffmann, A.A., Sørensen, J.G., Loeschcke, V., 2003. Adaptation of *Drosophila* to temperature extremes: bringing together quantitative and molecular approaches. *Journal of Thermal Biology* 28, 175-216.

Hofmann, G., Somero, G., 1995. Evidence for protein damage at environmental temperatures: seasonal changes in levels of ubiquitin conjugates and hsp70 in the intertidal mussel *Mytilus trossulus*. *Journal of Experimental Biology* 198, 1509-1518.

Kampinga, H., Hageman, J., Vos, M., Kubota, H., Tanguay, R., Bruford, E., Cheetham, M., Chen, B., Hightower, L., 2009. Guidelines for the nomenclature of the human heat shock proteins. *Cell Stress and Chaperones* 14, 105-111.

Kang, S.G., Choi, K.S., Bulgakov, A.A., Kim, Y., Kim, S.Y., 2003. Enzyme-linked immunosorbent assay (ELISA) used in quantification of reproductive output in the pacific oyster, *Crassostrea gigas*, in Korea. *Journal of Experimental Marine Biology and Ecology* 282, 1-21.

Krivoruchko, A., Storey, K.B., 2010. Forever young. *Oxidative Medicine and Cellular Longevity* 3, 186-198.

Kültz, D., 2005. Molecular and evolutionary basis of the cellular stress response. *Annual Review of Physiology* 67, 225-257.

Lango-Reynoso, F., Chávez-Villalba, J., Cochard, J.C., Le Pennec, M., 2000. Oocyte size, a means to evaluate the gametogenic development of the Pacific oyster, *Crassostrea gigas* (Thunberg). *Journal of Shellfish Research* 190, 183-199.

Lejart, M., 2009. Etude du processus invasif de *Crassostrea gigas* en Bretagne: Etat des lieux, dynamique et conséquences écologiques. Université de Bretagne occidentale, Brest , France.

Li, Y., Qin, J.G., Abbott, C.A., Li, X., Benkendorff, K., 2007. Synergistic impacts of heat shock and spawning on the physiology and immune health of *Crassostrea gigas*: An explanation for summer mortality in Pacific oysters. *American Journal of Physiology: Regulatory, Integrative and Comparative Physiology*.

Lubet, P., 1959. Recherches sur le cycle sexuel et l'émission des gametes chez les Mytilides et les Pectinides (Mollusques, Bivalves). *Revue des Travaux de l'Office des Pêches Maritimes* 23, 389–545.

Lubet, P., Herlin-Houtteville, P., Matthieu, M., 1976. La lignée germinale des mollusques pélécypodes. Origine et évolution. *Bull Soc Zool France* 101, 22–27.

Mann, R., 1979. Some biochemical and physiological aspects of growth and gametogenesis in *Crassostrea gigas* and *Ostrea edulis* grown at sustained elevated temperatures. *Journal of the Marine Biological Association of the United Kingdom* 59, 95-110.

Mao, Y., Zhou, Y., Yang, H., Wang, R., 2006. Seasonal variation in metabolism of cultured Pacific oyster, *Crassostrea gigas*. *Aquaculture* 253, 322-333.

Martin, M., Osborn, K.E., Billig, P., Glickstein, N., 1981. Toxicities of Ten Metals to *Crassostrea gigas* and *Mytilus edulis* Embryos and Cancer magister Larvae. *Marine Pollution Bulletin* 12, 305-308.

Meistertzheim, A.L., Tanguy, A., Moraga, D., Thebault, M.T., 2007. Identification of differentially expressed genes of the Pacific oyster *Crassostrea gigas* exposed to prolonged thermal stress. *FEBS Journal* 274, 6392-6402.

Meistertzheim, A.L., Lejart, M., Le Goïc, N., Thébault, M.T., 2009. Sex-, gametogenesis, and tidal height-related differences in levels of HSP70 and metallothioneins in the Pacific oyster *Crassostrea gigas*. *Comparative Biochemistry and Physiology. Part A, Molecular and Integrative Physiology* 152, 234-239.

Moraga, D., Meistertzheim, A.L., Tanguy-Royer, S., Boutet, I., Tanguy, A., Donval, A., 2001. Stress response in Cu2+ and Cd2+ exposed oysters (*Crassostrea gigas*): an immunohistochemical approach. *Comparative Biochemistry and Physiology. Toxicology and Pharmacology* 141, 151-156.

Mortensen, C.J., Choi, Y.H., Ing, N.H., Kraemer, D.C., Vogelsang, M.M., Hinrichs, K., 2010. Heat shock protein 70 gene expression in equine blastocysts after exposure of oocytes to high temperatures in vitro or in vivo after exercise of donor mares. *Theriogenology* 74, 374-383.

Newell, R.C., 1979. Biology of Intertidal Animals. Faversham, Kent, UK London.

Olesen, C., Møller, M., Byskov, A.G., 2004. Tesmin transcription is regulated differently during male and female meiosis. *Molecular Reproduction and Development* 67, 116-126.

Pörtner, H.O., 2002. Climate variations and the physiological basis of temperature dependent biogeography: systemic to molecular hierarchy of thermal tolerance in animals. *Comparative Biochemistry and Physiology. A, Comparative Physiology* 132, 739-761.

Quayle, O.B., 1969. Pacific oyster culture in British Columbia. *Journal of the Fisheries Research Board of Canada* a, 169-192.

Ringwood, A.H., Brouwer, M., 1995. Patterns of metallothionein expression in oyster embryos. *Marine Environmental Research* 39, 101-105.

Roberts, D.A., Hofmann, G.E., Somero, G.N., 1997. Heat-Shock Protein Expression in *Mytilus californianus*: Acclimatization (Seasonal and Tidal-Height Comparisons) and Acclimation Effects. *Biology Bulletin* 192, 309-320.

Roesijadi, G., 1992. Metallothioneins in metal regulation and toxicity in aquatic animals. *Aquatic Toxicology* 22, 81-114.

Roesijadi, G., Hansen, K.M., Unger, M.E., 1996. Cadmium-induced metallothionein expression during embryonic and early larval development of the mollusc *Crassostrea virginica*. *Toxicology and Applied Pharmacology* 140, 356-363.

Rupik, W., Jasik, K., Bembenek, J., Widłak, W., 2011. The expression patterns of heat shock genes and proteins and their role during vertebrate's development. *Comparative Biochemistry and Physiology - Part A: Molecular & Integrative Physiology* 159, 349-366.

Sanders, B.M., Hope, C., Pascoe, V.M., Martin, L.S., 1991. Characterization of the stress protein response in two species of *Collisella* limpets with different temperature tolerances. *Physiological Zoology*, 1471-1489.

Sconzo, G., Palla, F., Agueli, C., Spinelli, G., Giudice, G., Cascino, D., Geraci, F., 1999. Constitutive hsp70 Is Essential to Mitosis during Early Cleavage of *Paracentrotus lividus* Embryos: The Blockage of Constitutive hsp70 Impairs Mitosis. *Biochemical and Biophysical Research Communications* 260, 143-149.

Soletchnik, P., Lambert, C., Costil, K., 2005. Summer mortality of *Crassostrea gigas* (Thunberg) in relation to environmental rearing conditions. *Journal of Shellfish Research* 24, 197-207.

Somero, G.N., 2002. Thermal Physiology and Vertical Zonation of Intertidal Animals: Optima, Limits, and Costs of Living, pp. 780-789.

Somero, G.N., 2010. The physiology of climate change: how potentials for acclimatization and genetic adaptation will determine 'winners' and 'losers'. *Journal of Experimental Biology* 213, 912-920.

Tanguy, A., Moraga, D., 2001. Cloning and characterization of a gene coding for a novel metallothionein in the Pacific oyster *Crassostrea gigas* (CgMT2): a case of adaptive response to metal-induced stress? *Gene* 273, 123-130.

Thornalley, P.J., Vasak, M., 1985. Possible role for metallothionein in protection against radiation-induced oxidative stress. Kinetics and mechanism of its reaction with superoxide and hydroxyl radicals. *Biochimica et Biophysica Acta* 827, 36-44.

Tomanek, L., 2008. The importance of physiological limits in determining biogeographical range shifts due to global climate change: the heat-shock response. *Physiological and Biochemical Zoology* 81, 709-717.

Tomanek, L., Somero, G.N., 1999. Evolutionary and acclimation-induced variation in the heat-shock responses of congeneric marine snails (genus *Tegula*) from different thermal habitats: implications for limits of thermotolerance and biogeography. *Journal of Experimental Biology* 202, 2925-2936.

Tomanek, L., Helmuth, B., 2002. Physiological ecology of rocky intertidal organisms: a synergy of concepts. *Integrative and Comparative Biology* 42, 771-775.

Valavanidis, A., Vlahogianni, T., Dassenakis, M., Scoullos, M., 2006. Molecular biomarkers of oxidative stress in aquatic organisms in relation to toxic environmental pollutants. *Ecotoxicology and Environmental Safety* 64, 178-189.

Viarengo, A., Nott, J.A., 1993. Mechanisms of heavy metal cation homeostasis in marine invertebrates. *Comparative Biochemistry and Physiology, Part C: Toxicology and Pharmacology* 104, 355-372.

Wolff, W.J., Reise, K., 2002. Oyster imports as a vector for the introduction of alien species into Northern and Western European coastal waters. In: *Invasive Aquatic Species of Europe: Distribution, Impacts and Management* Leppäkoski, E., Gollasch, S., Olenin, S. (Eds.), Kluwer, Dordrecht pp. 193-205.

In: Heat Shock Proteins
Editor: Saad Usmani
ISBN: 978-1-62417-571-8

Chapter VII

Posttranslational Modification of Heat Shock Protein 25/27 (HSP25/27) by Methylglyoxal in Gastrointestinal Cancer

***Tomoko Oya-Ito*[*1], *Yuji Naito*[1], *Tomohisa Takagi*[1], *Osamu Handa*[1], *Keisuke Shima*[2] *and Toshikazu Yoshikawa*[1]**

[1]Department of Molecular Gastroenterology and Hepatology, Graduate School of Medical Science, Kyoto Prefectural University of Medicine, Japan

[2]Shimadzu Corporation

Abstract

Using a proteomics approach, we identified murine heat-shock protein 25 (HSP25) and human HSP27 as the major methylglyoxal adduct proteins in rat gastric carcinoma mucosal cell line and human colon cancer cell line, respectively. Furthermore, methylglyoxal modification of HSP27 was detected in ascending colon and rectum of patient with cancer, but not in normal subjects. There were no differences in the level of protein expression of HSP25 between in rat gastric mucosal cell line and in rat gastric carcinoma mucosal cell line, while the levels of lactate converted from methylglyoxal were increased in carcinoma mucosal cell line. Matrix-associated laser desorption/ionization (MALDI) mass spectrometry/mass (MS/MS) spectrometry analysis of the peptide fragments derived from methylglyoxal-modified HSP27 showed that Arg-188 is susceptible to modification. HSP27 inhibits apoptosis in both normal and cancer cells. The transfer of methylglyoxal-modified HSP27 was even more effective in preventing apoptotic cell death than that of native control HSP27. Additionally, methylglyoxal modification of HSP27 protected the cells against both the hydrogen peroxide- and cytochrome c-mediated caspase activations. Our results suggest that posttranslational modification of HSP27 by methylglyoxal may have important implications for epithelial cell injury in gastrointestinal cancer.

* email address: oya-ito@koto.kpu-m.ac.jp.

The molecular mechanisms underlying the posttranslational modification of proteins in gastrointestinal cancer are still unknown. Methylglyoxal is a reactive dicarbonyl compound produced from cellular glycolytic intermediates that reacts non-enzymatically with proteins to form products such as argpyrimidine at arginine residue. High levels of methylglyoxal are present in cultured cells, and the most biogenic methylglyoxal (99%) is involved in reversible and irreversible interactions *in vivo*. The posttranslational modification of proteins by methylglyoxal may contribute to aging, diabetes, and other disorders. The specific immuno-reactivity against the methylglyoxal-modified protein was observed in gastrointestinal carcinoma cell lines and in ascending colon and rectum of patients with cancer, but not in normal subjects [1]. By using a monoclonal antibody to methylglyoxal-modified proteins, we found that murine heat shock protein 25 (HSP25) and human HSP27 were the major adducted proteins. HSP27 (apparent molecular mass, 27 kDa) is the human homolog of rodent protein HSP25. Only in rat gastric carcinoma mucosal RGK-1 cells but not in rat gastric mucosal RGM-1 cells the immunoreactivity of a ~ 25 kDa protein was identified by an antibody to argpyrimidine, as indicated by arrow (figure 1C). RGK-1 cells are an N-methyl-N'-nitro-N-nitrosoguanidine (MNNG)-induced mutant of RGM-1 gastric epithelial cell line [2]. The expression levels of HSP25 showed no difference between in RGK-1 cells and RGM-1 cells. Our experiments, especially immunoprecipitation assays, showed HSP25 in rat intestinal epithelial cell line RIE and in rat gastric mucosal cell line RGM-1 were not modified by methylglyoxal, whereas HSP25 in rat gastric carcinoma mucosal cell line RGK-1 and HSP27 in human colon cancer cell line HT-29 and in ascending colon and rectum of patient with adenoma and advanced cancer were (figure 2). The levels of lactate converted from methylglyoxal were increased in carcinoma cell lines. It is difficult to measure intracellular levels of methylglyoxal directly because of its high reactivity. The glyoxalase system, comprising the metalloenzymes glyoxalase I and glyoxalase II, is an almost universal metabolic pathway involved in the detoxification of the glycolytic byproduct methylglyoxal to D-lactate. The results demonstrate that there is an increase in the flux of methylglyoxal metabolized to lactate via the glyoxalase pathway with high glucose concentrations in carcinoma cells.

Matrix-associated laser desorption/ionization (MALDI) mass spectrometry/mass spectrometry (MS/MS) analysis of peptide fragments identified Arg-75, Arg-79, Arg-89, Arg-94, Arg-127, Arg-136, Arg-140, Arg-188, and Lys-123 as methylglyoxal modification sites in HSP27 and in phosphorylated HSP27. The peptides identified by LC-MALDI analysis of enzymatic digests of methylglyoxal-modified HSP27 are listed in Table 1. Furthermore, we identified for the first time that argpyrimidine adducts are generated at Arg-136 and Arg-188 of methylglyoxal-modified HSP27. Arg-188 is thought to be necessary for its enhanced anti-apoptotic effects [3]. This modification process is essential to its repressing activity for cytochrome c-mediated caspase activation. Inhibition of methylglyoxal modification of HSP27 causes sensitization of the cells to anti-tumor drug-induced apoptosis. The C-terminal region of one subunit interacts with neighboring subunits by hydrophobic interaction and backbone hydrogen bonding. Our results showed that the Arg-188 residue at the C terminus is modified by methylglyoxal to form argpyrimidine. Methylglyoxal modification at the C terminus could be involved in hydrophobic interaction with neighboring subunits of HSP27 and the formation of large oligomers because argpyrimidine has the aromatic hydroxy group.

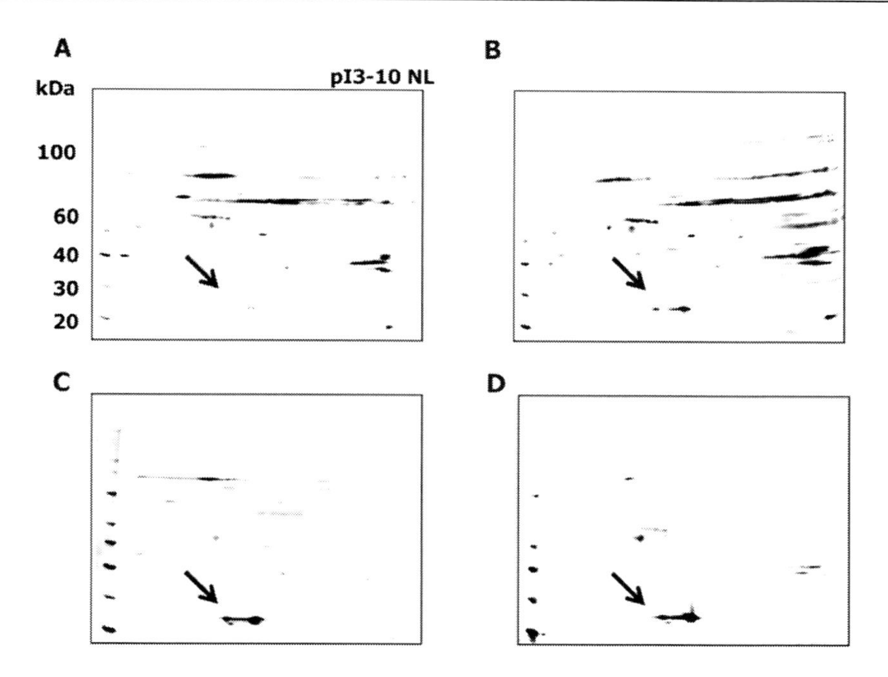

Figure 1. Two-dimensional electrophoresis and Western blot analysis of proteins obtained from RGM-1 cells and RGK-1 cells. Whole cell extracts from RGM-1 cells (A and C) and RGK-1 cells (B and D) were separated by two-dimensional electrophoresis. Methylglyoxal-modified proteins were identified by immunoblot analysis using an anti-argpyrimidine antibody (A and B). HSP25 (indicated by arrows) was identified by immunoblotting using an anti-HSP25 antibody (C and D).

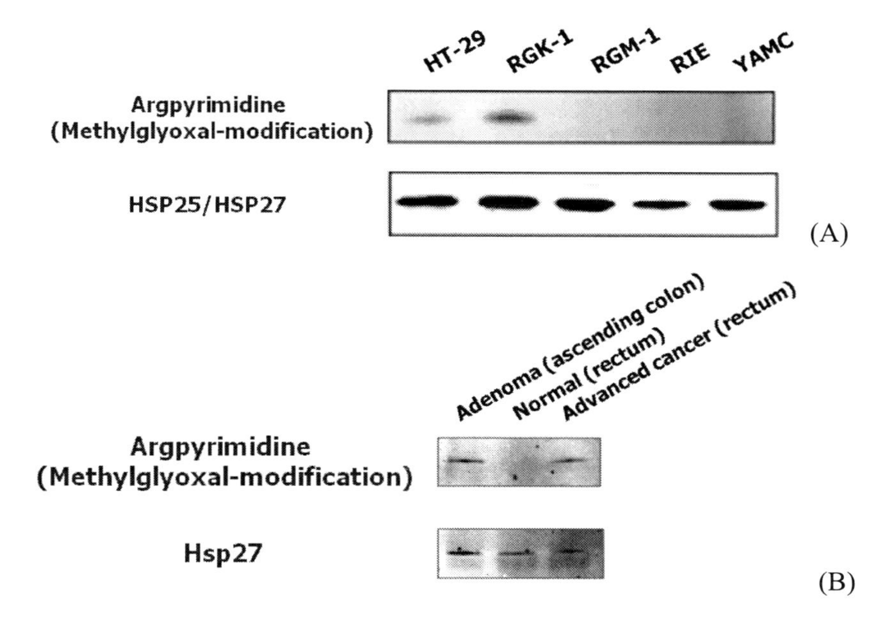

Figure 2. Detections of methylglyoxal-modified proteins and HSP25/HSP27. Methylglyoxal modification in immunoprecipitated HSP25/HSP27 from whole cell extracts (A) and from tissue extracts (B) were identified by immunoblot analysis using anti-argpyrimidine antibody. The immunoprecipitated HSP25/27 were verified by immunoblot analyses using anti-HSP25 antibody (RGM-1, RGK-1, RIE, and YAMC cells) and anti-HSP27 antibody (HT-29 cells and tissue of patients).

Table 1. Peptides identified by MALDI-MS and LC-MALDI-TOF MS/MS analyses of enzymatic digests of methylglyoxal-modified HSP27. Modification sites are underlined

Theoretical Mass	Start	End	Sequence	Modifications	Enzyme
1041.25	5	12	RVPFSLLR	5-Hydro-5-methylimidazolone (R)	Trypsin
1556.76	28	40	LFDQAFGLPRLPE	5-Hydro-5-methylimidazolone (R)	Trypsin+V8
2738.01	41	64	EWSQWLGGSSWPGYVRPLPPAAIE	5-Hydro-5-methylimidazolone (R)	Trypsin+V8
1570.75	65	79	SPAVAAPAYSRALSR	5-Hydro-5-methylimidazolone (R)	Trypsin+V8
2412.61	65	87	SPAVAAPAYSRALSRQLSSGVSE	2x 5-Hydro-5-methylimidazolone (R)	Trypsin+V8
1318.44	88	96	IRHTADRWR	2x 5-Hydro-5-methylimidazolone (R)	Trypsin+V8
2180.42	95	112	WRVSLDVNHFAPDELTVK	5-Hydro-5-methylimidazolone (R)	Trypsin
1769.91	113	127	TKDGVVEITGKHEER	Carboxyethyllysine (K)	Trypsin
2856.02	113	136	TKDGVVEITGKHEERQDEHGYISR	Carboxyethyllysine (K)	Trypsin
1540.63	115	127	DGVVEITGKHEER	Carboxyethyllysine (K)	Trypsin
2626.75	115	136	DGVVEITGKHEERQDEHGYISR	Carboxyethyllysine (K)	Trypsin
2680.79	115	136	DGVVEITGKHEERQDEHGYISR	5-Hydro-5-methylimidazolone (R); Carboxyethyllysine (K)	Trypsin
1709.73	124	136	HEERQDEHGYISR	5-Hydro-5-methylimidazolone (R)	Trypsin
2274.39	124	140	HEERQDEHGYISRCFTR	5-Hydro-5-methylimidazolone (R)	Trypsin
1722.84	128	140	QDEHGYISRCFTR	5-Hydro-5-methylimidazolone (R)	Trypsin
1350.5	131	140	HGYISRCFTR	5-Hydro-5-methylimidazolone (R)	Trypsin+V8
1376.54	131	140	HGYISRCFTR	Argpyrimidine (R)	Trypsin+V8
1532.72	131	141	HGYISRCFTRK	2x 5-Hydro-5-methylimidazolone (R)	Trypsin+V8
2883.17	172	198	LATQSNEITIPVTFESRAQLGGPEAAK	5-Hydro-5-methylimidazolone (R)	Trypsin
2883.17	172	198	LATQSNEITIPVTFESRAQLGGPEAAK	5-Hydro-5-methylimidazolone (R)	Trypsin
1238.35	187	198	SRAQLGGPEAAK	5-Hydro-5-methylimidazolone (R)	Trypsin+V8
1256.37	187	198	SRAQLGGPEAAK	Dihydroxyimidazolidine (R)	Trypsin+V8
1264.39	187	198	SRAQLGGPEAAK	Argpyrimidine (R)	Trypsin+V8

In normal cells, HSP27 participates in cytoskeletal, redox state and protein folding homeostasis. It is also involved in the protection of cells against stress such as apoptosis [4] and actin polymerization [5-7]. High levels of constitutive HSP27 expression have been detected in several cancer cells, particularly carcinomas [8, 9]. However, the mechanism regulating HSP27 action in cancer cells is still unknown. HSP27 is thought to increase the ability of some cancer cells to resist and to evade the apoptotic processes mediated by the immune system. In the current study, no noticeable differences in expression level of HSP25/HSP27 between noncancerous and cancerous cells were found. The transfer of methylglyoxal-modified HSP27 into RIE was even more effective in preventing apoptotic cell death than that of native control HSP27 [1]. The apoptotic RIE cells were assayed by flow cytometry using FITC-Annexin V binding and PI staining. Treatment with hydrogen peroxide resulted in an increase in FITC-Annexin V-positive cells. The introduction of native HSP27 prevented increases in the number of FITC-Annexin V positive cells when apoptosis was induced by hydrogen peroxide. The transfer of methylglyoxal-modified HSP27 into the cells dramatically reduced the number of FITC-Annexin V positive cells. These results suggest that the introduction of native HSP27 delays the induction of apoptosis, and the methylglyoxal modification of HSP27 completely inhibits the induction of apoptosis. Additionally, methylglyoxal modification of HSP27 protected the cells against the both hydrogen peroxide- and cytochrome *c*-mediated caspase activations, and the hydrogen peroxide-induced production of intracellular reactive oxygen species (ROS), as shown in figure 3. The introduction of HSP27 into RIE cells prevented both caspase-3 and caspase-9 activation. Methylglyoxal modification enhanced the inhibitory effect of HSP27 on caspase-3 ($p<0.01$ versus HSP27-introduced cells) and on caspase-9 ($p<0.001$ versus HSP27-introduced cells) activations. Hydrogen peroxide treatment caused an increase in the amount of intracellular ·OH of control RIE cells in a concentration-dependent manner. The introduction of methylglyoxal-modified HSP27 prevented ·OH generation in the cells treated with hydrogen peroxide compared with the introduction of native unmodified HSP27. These results indicate that methylglyoxal is a novel modulator of cell survival by directly incorporating with the specific protein. It has been reported that methylglyoxal modification of Arg-188 in HSP27 is essential for inhibiting cytochrome *c*-mediated caspase activation in 293T cells transfected with mutant of HSP27 (R188G, Arg-188 to glycine) [3]. We found that Arg-188 is susceptible to modification by methylglyoxal [1]. The large oligomers of HSP27, which bear chaperone-like activity, are the structural organization of HSP27 required for its anti-apoptotic activity [10]. In our study, numerous post-translational modifications of HSP25/HSP27 in a large number of cancer cells were found. Recombinant HSP27 and phosphorylated HSP27 (pHSP27) were polymerized during incubation with methylglyoxal *in vitro* [1]. Moreover, large methylglyoxal-modified HSP27 and pHSP27 oligomers were detected in tumorigenic cells. These results suggest that the polymerizations of HSP27 and pHSP27 are promoted by methylglyoxal modification. Sakamoto *et al.* showed that HSP27 proteins that can form large oligomers interact with cytochrome *c* to interfere with caspase-dependent apoptosis and that the mutant R188G of HSP27 cannot form the large oligomers [3]. Therefore, methylglyoxal modification may modulate the protective ability of HSP27 through the interaction of cytochrome *c*. Modification by methylglyoxal enhanced the chaperone function of HSP27, whereas phosphorylation of HSP27 entirely abolished its chaperone function [4].

Interestingly, chaperone activity of even pHSP27 was greatly facilitated by methylglyoxal modification. Nagaraj *et al.* showed that argpyrimidine modification may be necessary for the improvement of the chaperone function in HSP27 [11]. An increased molecular chaperone function in cancer cells is likely related to oligomerization of HSP27 by methylglyoxal modification. In our previous study, we found that HSP27 shows increased anti-apoptotic effects in lens epithelial cells after modification by methylglyoxal, which is due to multiple mechanisms including the enhancement of chaperone function, and inhibition of caspase activity [4, 12, 13]. The transfer of methylglyoxal-modified HSP27 significantly inhibits staurosporine-induced apoptotic cell death and ROS production in a human lens epithelial cell line [4]. While methylglyoxal modification of HSP27 can maintain transparency and protect against apoptosis in lens cells which has a slow turnover rate, this modification may be involved in the pathogenesis of cancer in the gastrointestinal epithelium which has a rapid turnover rate. Methylglyoxal can alter the balance between induction and inhibition apoptosis. High concentrations of methylglyoxal act as a pro-apoptotic modulator by activating c-Jun N-terminal kinase and caspase proteases [14]. Conversely, our results indicated that endogenous methylglyoxal prevents cells from undergoing apoptosis by maintaining HSP27 anti-apoptotic activity. Methylglyoxal-mediated apoptotic effects may also depend on the expression levels of methylglyoxal-modified protein, including HSP27. Overall, our findings have important implications for understanding the pathogenesis of epithelial cell injury in gastrointestinal cancer.

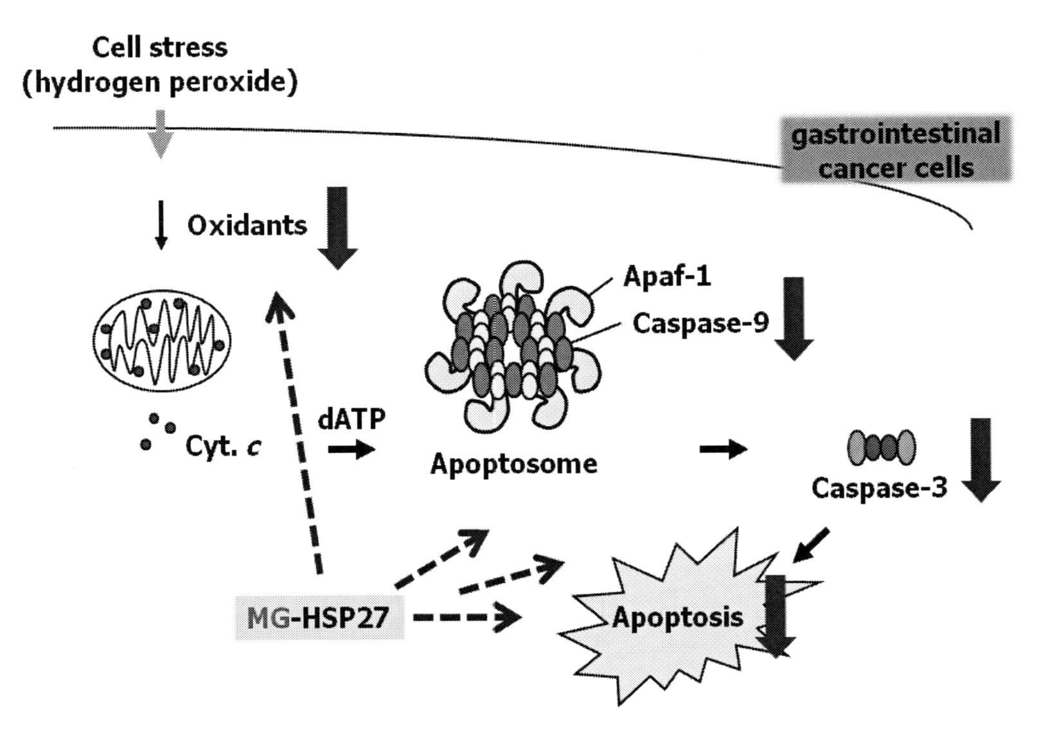

Figure 3. A proposed mechanism for prevention of apoptosis by HSP27. Methylglyoxal modification in gastrointestinal cancer cells promotes HSP27 properties. MG, methylglyoxal; Cyt. *c*, cytochrome *c*.

References

[1] Oya-Ito, T., Naito, Y., Takagi, T., Handa, O., Matsui, H., Yamada, M., Shima, K., and Yoshikawa, T. (2011) *Biochim Biophys Acta* 1812, 769-781

[2] Shimokawa, O., Matsui, H., Nagano, Y., Kaneko, T., Shibahara, T., Nakahara, A., Hyodo, I., Yanaka, A., Majima, H. J., Nakamura, Y., and Matsuzaki, Y. (2008) *In Vitro Cell Dev Biol Anim* 44, 26-30

[3] Sakamoto, H., Mashima, T., Yamamoto, K., and Tsuruo, T. (2002) *J Biol Chem* 277, 45770-45775

[4] Oya-Ito, T., Liu, B. F., and Nagaraj, R. H. (2006) *J Cell Biochem* 99, 279-291

[5] Negre-Aminou, P., van Leeuwen, R. E., van Thiel, G. C., van den, I. P., de Jong, W. W., Quinlan, R. A., and Cohen, L. H. (2002) *Biochem Pharmacol* 64, 1483-1491

[6] Huot, J., Houle, F., Rousseau, S., Deschesnes, R. G., Shah, G. M., and Landry, J. (1998) *J Cell Biol* 143, 1361-1373

[7] Lee, J. H., Sun, D., Cho, K. J., Kim, M. S., Hong, M. H., Kim, I. K., and Lee, J. S. (2007) *J Cancer Res Clin Oncol* 133, 37-46

[8] Ciocca, D. R., and Calderwood, S. K. (2005) *Cell Stress Chaperones* 10, 86-103

[9] Calderwood, S. K., Khaleque, M. A., Sawyer, D. B., and Ciocca, D. R. (2006) *Trends Biochem Sci* 31, 164-172

[10] Bruey, J. M., Paul, C., Fromentin, A., Hilpert, S., Arrigo, A. P., Solary, E., and Garrido, C. (2000) *Oncogene* 19, 4855-4863

[11] Nagaraj, R. H., Panda, A. K., Shanthakumar, S., Santhoshkumar, P., Pasupuleti, N., Wang, B., and Biswas A. (2012) *PLoS One*7, e30257.
[12] Nagaraj, R. H., Oya-Ito, T., Padayatti, P. S., Kumar, R., Mehta, S., West, K., Levison, B., Sun, J., Crabb, J. W., and Padival, A. K. (2003) *Biochemistry* 42, 10746-10755
[13] Biswas, A., Miller, A., Oya-Ito, T., Santhoshkumar, P., Bhat, M., and Nagaraj, R. H. (2006) *Biochemistry* 45, 4569-4577
[14] Kang, Y., Edwards, L. G., Thornalley, P. J. (1996) *Leuk Res* 20, 397-405.

Index

A

B

D

E

F

G

H

I

N

O

P

Q

R

S

T

U